SATAN, CANTOR, AND INFINITY

SATAN, CANTOR, AND INFINITY

AND OTHER MIND-BOGGLING PUZZLES

by

RAYMOND SMULLYAN

ALFRED A. KNOPF, NEW YORK
1992

"The Lie Detective," "When I Was a Boy," "The Abduction of
Annabelle," "How Kazir Won His Wife," "A Plague of Lies," and "On
the Other Hand" were originally published in *The Sciences.*

Library of Congress Cataloging-in-Publication Data
Smullyan, Raymond M.
Satan, Cantor, and infinity, and other mind-boggling puzzles /
Raymond Smullyan.—1st ed.
p. cm.
ISBN 0-679-40688-3
1. Mathematical recreations. 2. Gödel's theorem. I. Title.
QA95.S5 1992 92-52893 CIP
793.7′4—dc20

Manufactured in the United States of America

First Edition

Once again, I would like to thank my editor, Ann Close, and the
production editor, Melvin Rosenthal, for all their expert assistance. My
thanks go also to Matina Billias for her excellent secretarial help.

Contents

Preface

Few things have stirred the imagination as much as Infinity! It has all sorts of curious properties which at first seem paradoxical, but then turn out not to be. As such, it provides ideal material for a puzzle book.

As with my earlier puzzle books, this one starts out with new puzzles about truth-tellers and liars ("knights" and "knaves"), but I have added a remarkable character known as the Sorcerer, who is considered a magician by those around him, although he is really a logician who uses logic so cleverly that it seems like magic to those not in the know. After many exhibitions of his "logical sorcery," he escorts us through a host of unusual adventures, including a visit to an island where intelligent robots create other robots and endow them with enough intelligence to create other intelligent robots, who in turn create other intelligent robots, and so forth, ad infinitum. Then, after some special puzzles related to Gödel's famous theorem and some curious paradoxes about probability, time, and change, the Sorcerer gives us a guided tour of Infinity, explaining the pioneering discoveries of the great mathematician Georg Cantor, who was the first to put the subject on a logically sound basis. The Sorcerer, in his typically humorous fashion, ends up with a delightful tale of how Satan himself is outwitted by a clever student of Cantor's.

On the serious side, it must be wondered at that the whole fascinating subject of Infinity is so little known to the general public! Why isn't it taught in high schools? It is no harder to understand than algebra or geometry, and it is so rewarding! The last few chapters of this book provide an inviting (and easy) introduction to the subject. Even a neophyte can understand the nature of Infinity, Cantor's amazing contribution, and an account of what may be the greatest mathematical problem of all time—which remains unsolved to this very day!

A NOTE TO THE READER

The several parts of the book do not have to be read in the order in which they appear. Thus, the reader primarily interested in Infinity can read Parts VI and VII quite independently of the rest of the book. Parts III and IV likewise form a separate unit, and Parts I, II, and V can each be read independently.

All that a reader starting with later chapters needs to know is that the principal characters are the Sorcerer and his two students, Annabelle and Alexander.

PART I

LOGICAL SORCERY

1

THE LIE DETECTIVE

WITH A TWINGE of apprehension such as he had never felt before, an anthropologist named Abercrombie stepped onto the Island of Knights and Knaves. He knew that this island was populated by most perplexing people: knights, who make only true statements, and knaves, who make only false ones. "How," Abercrombie wondered, "am I ever to learn anything about this island if I can't tell who is lying and who is telling the truth?"

Abercrombie knew that before he could find out anything he would have to make one friend, someone whom he could always trust to tell him the truth. So when he came upon the first group of natives, three people, presumably named Arthur, Bernard, and Charles, Abercrombie thought to himself, "This is my chance to find a knight for myself." Abercrombie first asked Arthur, "Are Bernard and Charles both knights?" Arthur replied, "Yes." Abercrombie then asked: "Is Bernard a knight?" To his great surprise, Arthur answered: "No." Is Charles a knight or a knave?

Abercrombie knew that he must first determine what type (knight or knave) Arthur and Bernard are. Arthur is obviously a knave, since no knight would claim that Bernard and Charles

3

are both knights yet deny that Bernard is a knight. Therefore both of Arthur's answers were lies. Since he denied that Bernard is a knight, Bernard really is a knight. Since he affirmed that both Bernard and Charles are knights, it is false that they are both knights; at least one of them must be a knave. But Bernard is not a knave (as we have proved), therefore, Charles must be a knave.

Abercrombie was then informed by the one of the three he knew to be a knight that the island had a sorcerer.

"Oh, good!" Abercrombie exclaimed. "We anthropologists are particularly interested in sorcerers, witch doctors, medicine men, shamans, and the like. Where do I find him?"

"You must ask the King," came the reply.

Well, the anthropologist was able to obtain an audience with the King and told him that he wished to meet the Sorcerer.

"Oh, you can't do that," said the King, "unless you first meet his apprentice. If the Sorcerer's Apprentice approves of you, then he will allow you to meet his master; if he doesn't, then he won't."

"The Sorcerer has an apprentice?" asked the anthropologist.

"He certainly does!" replied the King. "There is a famous musical composition about him—I believe the composer was Dukas. Anyway, if you wish to meet the Sorcerer's Apprentice, he is now at his home, which is the third house on Palm Grove. At the moment he is entertaining two guests. If, when you arrive, you can deduce which of the three present is the Sorcerer's Apprentice, I believe that will impress him sufficiently that he will allow you to meet the Sorcerer. Good luck!"

A short walk brought the anthropologist to the house. When he entered, there were indeed three people present.

"Which of you is the Sorcerer's Apprentice?" asked Abercrombie.

"I am," replied one.

"I am the Sorcerer's Apprentice!" cried a second.

But the third remained silent.

"Can *you* tell me anything?" Abercrombie asked.

"It's funny," answered the third one with a sly smile. "At most, only one of the three of us ever tells the truth!"

Can it be deduced which of the three is the Sorcerer's Apprentice?

Here is how the anthropologist reasoned: If the third one is a knave, then his statement is false, which means that at least two of them are knights. But the first and second guests cannot both be knights, since their statements conflict. Therefore the third guest cannot be a knave; he must be a knight. This means his statement is true: he is the only knight present. Since the other two are knaves, their claims are both false; hence neither of them is really the Sorcerer's Apprentice. Ergo, it must be the third one.

The Apprentice was delighted with Abercrombie's reasoning and informed him that he could meet the Sorcerer.

"He is now upstairs in the tower conferring with the island Astrologer," said the Apprentice. "You may go up and interview them if you like, but please knock before entering."

The anthropologist went upstairs, knocked on the door, and was bidden to enter. When he did, he saw two very curious individuals, one wearing a green conical hat and the other a blue one. He could not tell from their appearance which was the Astrologer and which was the Sorcerer. After introducing himself he asked, "Is the Sorcerer a knight?" The one in the blue hat answered the question (he answered either yes or no), and the anthropologist was able to deduce which was the Sorcerer. Which one was the Sorcerer?

This type of puzzle is very different from the preceding two; it is a metapuzzle, *for the reader is not given all the pieces of the puzzle but information about the process of solving the puzzle. The reader, in other words, is not told what answer was given by the man in the blue hat; nevertheless he is told that the anthropologist could solve the puzzle after getting an answer; it is* this *information that is vital.*

Let us see how this sort of puzzle works: Suppose the man in the blue hat had answered yes; could the anthropologist then have known which one was the Sorcerer? Certainly not; the man who answered could be a knight, in which case all that would follow is that the Sorcerer is a knight; but the two might both be knights and either one could be the Sorcerer. Or again, the man who answered could be a knave, in which case the Sorcerer is a knave and could be either of the two (as far as the anthropologist could know). So if yes was the answer, the anthropologist could not have deduced which man was the Sorcerer. But we are given that the anthropologist did *deduce which was the Sorcerer; therefore, he must have gotten the answer no.*

Now we know that the speaker (the one with the blue hat) answered no. If the speaker is a knight, his answer was truthful; hence the Sorcerer is really not a knight. And since the speaker is a knight, then he is not the Sorcerer. On the other hand, if the speaker is a knave, his answer was a lie, which means that the Sorcerer must be a knight; hence, again, the speaker cannot be the Sorcerer. This proves that a no answer indicates that the speaker is not the Sorcerer, regardless of whether the answer was the truth or a lie. And so the man in the blue hat must be the Astrologer and the one in the green hat must be the Sorcerer.

In summary, a no answer proves that the man in the green hat is the Sorcerer, whereas a yes answer proves nothing at all.

Since the anthropologist was able to deduce the Sorcerer's identity, he must have gotten a no answer and deduced that the man in the green hat was the Sorcerer.

While the anthropologist had deduced which one was the Sorcerer, he did not yet know whether the Sorcerer was a knight or a knave. With one more question he discovered that the Sorcerer was a knight and the Astrologer a knave. And the Astrologer, a bit embarrassed, arose and left, saying: "According to the planetary configurations, I should be home now."

"Those astrologers," said the Sorcerer with a laugh, "humbugs every one of them. Now, with me it is different; my sorcery is real."

"To tell you the truth," said Abercrombie, "I am rather skeptical about the existence of magic."

"Oh, you don't understand," said the Sorcerer. "My sorcery doesn't use magic—though it seems to those around here that it does. My sorcery involves the clever use of logic. With my logic I am constantly fooling these fellows."

"Can you give me an example?" asked Abercrombie.

"Why certainly. Are you a betting man?"

"Occasionally," replied Abercrombie with some caution.

"Oh, it doesn't have to be a large bet; we'll wager just one copper coin. I will ask you a question to which you must answer yes or no. Even though the question has a definite correct answer, I'll wager that you will be powerless to give it. Anybody but you might be able to give the correct answer, but you cannot. In fact it will be *logically impossible* for you to give the correct answer, even though the question has one. Doesn't this sound like sorcery?"

"It certainly does," replied Abercrombie, who was enormously intrigued. "I'll take the bet—mainly because I'm so curious. What question do you have in mind?"

The Sorcerer then asked Abercrombie a yes or no question

that definitely had one and only one correct answer. And Abercrombie soon realized, to his surprise and amusement, that the Sorcerer was right. It was logically impossible for him to give the correct answer, even though he knew what it was.

Can you guess what question the Sorcerer asked?

The Sorcerer asked Abercrombie, "Will you answer no to this question?" If Abercrombie were to answer no, he would be denying that no is his answer; hence he would be wrong. If he were to answer yes, then he would be affirming that no is his answer; hence he would again be wrong. Thus it is logically impossible for Abercrombie to give the correct answer.

Clever, these sorcerers!

INSPECTOR CRAIG PAYS A VISIT

A few weeks after the anthropologist set sail for home, my friend Inspector Craig, of Scotland Yard, paid a visit to the island. On the first evening of this particular visit, Craig was invited to dine with the Chief Justice, a knight.

"Ah, yes," said the judge proudly, "yesterday I caught a knave at a trial and sentenced him to three months for perjury. One shouldn't lie under oath."

"Are you implying that it is all right to lie when *not* under oath?" asked Craig.

"No, no," cried the judge. "One should never lie at all—but especially under oath."

"Tell me what happened," said Craig, who took a great interest in such matters.

"Why, there were two defendants, named Barab and Zork. I knew that Barab personally disliked Zork, but that doesn't excuse his lying about him."

"What lie did he tell?" asked Craig.

"He claimed that a few minutes before the trial, he heard Zork confide to a friend: 'I lied yesterday.' "

"So what?" asked Craig.

"So I obviously convicted Barab for lying."

"How do you know he was lying?"

"Oh, come now," said the judge with some irritation. "I thought you were a good logician. Obviously, Zork could never have said that he lied yesterday, because a knight would never falsely state that he lied yesterday and a knave would never truthfully admit that he lied yesterday. Therefore Barab clearly lied when he said that Zork made that statement."

"Not necessarily so," replied Craig. "You should brush up on your own logic and, of more immediate concern, you should have Barab released right away, since you have sentenced him without just cause."

Subsequent investigation revealed that Craig was right. The judge made a rather natural error in his reasoning, but an error nevertheless. What is the error?

Subsequent investigation revealed that Barab was in fact a knight and was telling the truth: Zork had in fact made that strange statement. How could Zork have claimed that he lied yesterday? Well, it turned out that Zork had laryngitis the day before and, hence, made no statements at all on that day. So Zork was a knave who lied when he claimed to have lied the day before; he had really been silent.

The next day Craig was asked to preside as judge at a trial concerning a stolen watch. The defendant was named Gary. Inspector Craig was not interested in finding out whether Gary was a knight or a knave; all he wanted to know was whether Gary did or did not steal the watch. Here is a transcript of the trial:

Craig: Is it true that sometime after the robbery, you claimed that you were not the one who stole the watch?

Gary: Yes.

Craig: Did you ever claim that you *were* the one who stole the watch?

Gary then answered (yes or no), and Craig knew whether or not he was innocent of the theft. Did Gary steal the watch?

This puzzle is another example of a metapuzzle. Suppose Gary had answered yes to Craig's second question. Then it would be obvious that Gary is a knave, because a knight could never claim to have made two contradictory claims. Since Gary is a knave (still under the supposition that he answered yes), then both his answers were lies, which means that he never claimed that he was not the thief, nor has he ever claimed that he was the thief, and so Craig could have no grounds for knowing whether Gary was innocent or guilty. But Craig did know. Hence Gary's second answer could not have been yes; it must have been no.

Now that we know that Gary's second answer was no, we can determine his innocence or guilt. Gary is either a knight or a knave. Suppose he is a knight. Then both his answers were truthful, which means he did claim once that he was not the thief, but he never claimed that he was the thief. Since he once claimed that he was not the thief and he is a knight, then he is innocent. On the other hand, suppose he is a knave. Then both his answers were lies, which means that he never did claim he was not the thief, but he did claim he was the thief. Then, being a knave and claiming that he was the thief means that in reality he was not the thief, and so again he is innocent. This proves that regardless of whether Gary is a knight or a knave, he is innocent of the crime.

2

WHEN I WAS A BOY

WHEN THE ANTHROPOLOGIST Abercrombie returned from the Island of Knights and Knaves, he called in the press to relate his adventures with the Sorcerer's Apprentice and his audience with the Sorcerer himself. A reporter named Bill Ryan was so intrigued that he decided to visit the island and interview the Sorcerer. After setting sail one wintry day from Baltimore, he arrived on the island and tracked down the mysterious Sorcerer in his mountain castle.

"Tell me," said Ryan, to the Sorcerer, pencil and notebook in hand, "when did you first get interested in logic?"—for the Sorcerer's "magic," as Abercrombie had discovered, consisted of nothing more than the clever use of logic.

"It started when I was a boy," came the reply. "My uncle told me about a mythical Island of Knights and Knaves. (I now have reason to believe that he knew such an island actually existed, but wanted to test me before raising my hopes about traveling there.) Anyway, he first told me about a shipwrecked traveler who came across three island natives named Anthony, Bertrand, and Clive. He asked one: 'Are you a knight or a knave?' Anthony answered, but in a foreign tongue. The traveler then asked Bertrand what Anthony had said. Bertrand replied: 'Anthony said that he is a knave.' But Clive chimed in: 'Don't believe Bertrand; he is lying.'

"The traveler (who I now believe was really my uncle) was perplexed. And then it suddenly dawned on him which type (knight or knave) Clive was. But my uncle would not tell me the answer. I had to figure it out myself. Can you guess whether Clive was a knight or a knave?"

Ryan was stumped, so the Sorcerer explained.

"I looked up at my uncle," recalled the Sorcerer, "and said that no inhabitant of a knight-knave island could possibly claim that he was a knave, because a knight would never lie and claim to be a knave, and a knave would never truthfully admit to being a knave. I reasoned then that Bertrand clearly lied when he said that Anthony had claimed to be a knave, and hence Clive had told the truth when he said that Bertrand lied. Therefore, Bertrand was a knave and Clive was a knight."

"Now that I live on this island, I often remember my uncle's tales as I try to unmask knights and knaves."

"After all these years on the island, don't you know who is a knight and who is a knave?" asked Ryan.

"Why, not two weeks ago," said the Sorcerer, "I was walking on the beach and came across someone I didn't know. There were no visitors on the island at the time, so I knew he had to be a native. But I had no idea whether he was a knight or a knave. He mumbled a few words as we passed. I thought for a moment, and then shouted after him: 'If you hadn't made that statement, I could have believed it! Before you spoke, I had no idea whether you were a knight or a knave, nor did I have the slightest idea of whether what you just said was true or false. Now I know that your statement is false and that you must be a knave.' "

"What could he possibly have said that made you react like that?" asked Ryan.

"What do you think he said?" challenged the Sorcerer.

"Well, he might have said 'Two plus two equals five,'" replied Ryan. "Wouldn't that be enough to convince you that he was a knave?"

"Of course it would!" exclaimed the Sorcerer. "But I see you don't understand the problem! Had he said *that*, I would have known that he was a knave only because I knew before he spoke that 'Two plus two equals five' is a false statement. But I told you that it was only *after* he made the statement that I could deduce it was false—false by virtue of the very fact that he had made it. Now can you supply such a statement?"

Again, the reporter was baffled.

"When I first met the native," explained the Sorcerer, "I had no idea whether he was a knight or a knave, nor whether he was married. But then he said, 'I am a married knave.' Obviously a knight could never claim to be a married knave (or any other kind of knave, for that matter), and so the native was certainly a knave. Therefore, his statement was false. He was not really a married knave, so he must have been a single knave. After he spoke, I knew two things about him that I hadn't known before—that he was a knave, and that he was not married."

"Now, just a minute!" exclaimed Ryan. "I can't accept your solution as valid. Indeed, I can't see how that incident could ever have taken place. And I am deeply disturbed that you, a knight, could have ever told me this false story. Or perhaps you are a knave and everything you have told me today is untrue."

"Why do you say that the story is false?" asked the Sorcerer in surprise.

"Because *no* inhabitant of this island could possibly claim to be a knave. If he claims to be a married knave, then he is certainly claiming to be a knave, which cannot be done, as your uncle (if he really existed) pointed out. Your story doesn't hold water."

"Not so fast, young man," said the Sorcerer. "You have just committed a rather common fallacy. Let me ask you a question: Suppose a person claims to know both French and German. Is he necessarily claiming that he knows French?"

"Well, of course," replied Ryan. "What a silly question."

"Not so silly, if you stop and think about it. Let me put the question this way: A person has just told you that he knows both French and German. Then you ask him, 'Do you know French?' Would he necessarily claim that he does?"

"Of course," replied Ryan. "Why wouldn't he?"

"Ah, there's where you're wrong. He might and he might not—he might even *deny* that he knows French."

Ryan scratched his head as the Sorcerer explained.

"If a truthful person claimed to know both French and German, then he would of course also claim to know French. But with a liar, it's different; if he happened to know French and not German, then he would make the false claim that he knows both French and German. Then if you asked whether he knows French, he would lie and say no. Similarly, a native islander could claim to be a married knave, yet deny that he is a knave. This is a curious fact about the logic of lying and truth-telling that takes some getting used to."

It didn't take Ryan very long to realize that the Sorcerer was right.

"On another occasion," said the Sorcerer, "I came across a native who made a statement from which I could deduce that the statement must be true, but I did not know the truth of the statement before it was made, nor did I have any prior knowledge that he was a knight. Now, if you really understand what I just taught you, Ryan, you will know what the native could have said."

Ryan thought a bit and then answered boldly: "The native could have said, 'I am not a married knight.' Clearly a knave couldn't say that (because a knave is indeed not a married knight). Since he is a knight his statement is true and so he must be an unmarried knight."

"Suppose," said the Sorcerer, "that the native had instead said, 'I am an unmarried knight.' Could you deduce whether he is a knight or a knave or whether he is married?"

Ryan was ready to answer yes, when he caught himself. "No, I could not. Any knave could claim to be an unmarried knight (since he is not an unmarried knight) and any unmarried knight could also make that claim. All we can deduce from his claim is that if he is a knight then he is unmarried, which doesn't tell us much."

"Whew," said Ryan. "I almost confused the two statements: 'I am an unmarried knight' and 'I am not a married knight.' They really say quite different things."

"You are progressing well, much better than I would have thought when you first walked in. But I'm afraid," said the Sorcerer, as he looked up at his great antique clock, "that I must go off to a trial. It promises to be a very interesting one. The presiding judge is a former student of the remarkable logician Inspector Craig. Would you like to come with me?"

"With pleasure," replied Ryan.

The two descended from the Sorcerer's tower and wended their way to the courthouse. Along the way, the Sorcerer told Ryan what was already known about the case.

"The case concerns a stolen horse. There are four suspects—Andrew, Bruce, Clayton, and Edward. The authorities know for sure that one and only one of these four is the thief. The first three have already been found and put in custody, but Edward cannot be found anywhere. The trial will have to proceed without him."

Ryan and the Sorcerer arrived at the courthouse none too soon; the trial commenced just as they were seated. And, since almost everyone on the island was interested in the case, the house was filled to capacity.

First the judge pounded his gavel and asked a highly relevant question: "Who stole the horse?" He got the following replies:

Andrew: Bruce stole the horse.

Bruce: Clayton stole the horse.

Clayton: It was Edward who stole the horse.

Then, quite unexpectedly, one of the three defendants said, "The other two are lying."

The judge thought for a bit, then pointed to one of the three and said, "Obviously, *you* didn't steal the horse, so you may leave the court."

The acquitted man was happy to comply, and so only two defendants were left on trial.

The judge then asked one of the remaining two whether the other was a knight, and, after receiving an answer (yes or no), knew who stole the horse. What did the judge decide?

First we must determine whom the judge immediately acquitted. Suppose it was Andrew. If Andrew is a knight, then Bruce must be guilty and Andrew innocent. If Andrew is a knave, then it is false that Bruce and Clayton both lied; at least one of them told the truth. This means that either Clayton is guilty (if Bruce told the truth), or Edward is guilty (if Clayton told the truth); in either case, Andrew would be innocent. And so if it was Andrew who made the second statement, he is innocent, regardless of whether he is a knight or a knave. The judge, of course, would have realized this and acquitted him.

However, if either Bruce or Clayton made the second statement, the judge could not have found grounds to acquit anyone. If Bruce spoke, the judge could only tell that either Bruce is a knight and Clayton is guilty, or Bruce is a knave and either Bruce or Edward is guilty. If it was Clayton who spoke, then the judge could only tell that either Clayton is a knight and Edward is guilty, or Clayton is a knave and either Bruce or Clayton is guilty. Since the judge did make an acquittal, it must have been Andrew who spoke and was acquitted.

Thus the remaining defendants were Bruce and Clayton. One of the two, by answering the judge's last question, either claimed that the other defendant was a knight, or that the other was a knave. If the former, then the two defendants are the same type (both knights or both knaves); if the latter then the two are different types. Suppose the latter. Then Clayton may be a knight and Bruce a knave, in which case Edward is guilty (because Clayton said he was), or Bruce may be a knight and Clayton a knave, in which case Clayton is guilty. However, the judge couldn't have known which, and hence couldn't make a conviction. Therefore, one of the two must have claimed that the other was a knight (by answering yes to the judge's question). The judge then knew that they are of the same type. They can't both be knights (since their accusations conflict), so they are both knaves and their accusations are both false. Neither Clayton nor Edward stole the horse. Andrew, as we know, has already been acquitted. So it was Bruce who stole the horse.

"Speaking of horses," said the Sorcerer to Ryan as they were leaving the courthouse after the trial, "I must tell you of an extremely funny incident that happened many years ago when

I was living in a small town in another land. An individual named Archibald sold a horse to an individual named Benjamin. I knew both men, and was quite curious to know which of them had made the shrewder bargain. First I asked Benjamin how much he had paid for the horse. He named a surprisingly low figure. Then I turned to Archibald and asked, 'Why did you sell such a magnificent animal for so little?' Archie replied, 'He got no bargain; the horse is lame.' Next I asked Benjamin, 'How come you paid so much for a lame animal?' Benjamin replied, 'He isn't really lame. You see there is a nail sticking in his foot, which makes him limp and *appear* lame. Archibald, no doubt, thought he *was* lame, and that's why he sold him for so little. But when I take the horse away, I'll pull the nail out and he'll be as good as new.'

"Turning to Archibald, I said, 'Aha! I see he got the best of you. The horse isn't really lame.' But Archie replied, 'No, no; the horse really *is* lame. I simply stuck a nail in his foot to make Benjamin think that *it* was causing the horse to limp. But when he takes the nail out, he'll see that the horse limps as much as ever.'

"At this, I said to Benjamin, 'So, he really cheated you. The horse *is* lame; Archibald deliberately put the nail in the horse's foot just to mislead you.' Upon which Benjamin replied, 'I considered that possibility, and that's why I paid him in counterfeit money.' "

3

THE ABDUCTION OF
ANNABELLE

ONE NOVEMBER NIGHT, on an uncharted atoll a few hundred leagues distant from the Island of Knights and Knaves, Princess Annabelle, the King's younger daughter, was kidnapped. Rumor had it that she had been taken by boat to the Island of Knights and Knaves, but no one really knew whether it was true. Annabelle's suitor, a strapping youth named Alexander, immediately set sail for the island, hoping to find out if she was imprisoned there. He assumed (and rightly so) that if she had been taken to the island, she would still be on it. He also rightly reasoned that the island's Witch Doctor would know if Annabelle was still being held captive. The only trouble was that Alexander did not know whether the witch doctor was a knight or a knave!

Alexander arrived safely on the island, sought out the Witch Doctor, and asked, "Is Princess Annabelle on this island?" The Witch Doctor answered either yes or no. Then the suitor asked, "Have you *seen* Princess Annabelle on this island?" The Witch Doctor answered either yes or no, and the suitor then knew whether Annabelle was on the island. Was she?

Suppose that both the Witch Doctor's answers were yes. Then it could be that the Witch Doctor is a knight and that Annabelle

19

is on the island, or it could be that he is a knave and Annabelle is not there—Alexander could have no way of knowing. If both the Witch Doctor's answers were no, then it could be that the Witch Doctor is a knave and Annabelle is there, or it could be that he is a knight and Annabelle is not there—again, Alexander would be unable to decide. But since he does know whether Annabelle is on the island, he must have gotten one yes answer and one no answer.

Let us see what would happen if the first answer was yes and the second answer no. If the Witch Doctor were a knave, then both answers would be lies, meaning that Annabelle was never there, but that the Witch Doctor had seen her there, which is not possible. Therefore, the Witch Doctor must be a knight and Annabelle must be on the island (though the Witch Doctor never saw her there).

On the other hand, suppose the first answer was no and the second answer yes. If the Witch Doctor were a knight, then we would again have the impossible situation that Annabelle was never on the island, yet he had seen her there. Therefore the Witch Doctor must be a knave, and Annabelle must be on the island.

Of course, we have no way of knowing what answers the Witch Doctor actually gave (other than that one was yes and one was no), nor can we tell whether the Witch Doctor is a knight or a knave (though Alexander knew). But we have seen that only two sets of answers are possible (given that Alexander did know whether she was there), and in either case, Annabelle must be on the island.

Needless to say, Alexander was overjoyed to discover that his beloved Annabelle was on the island. The next step was to bargain for her release. With this in mind, he obtained an audience with the King, who was known to be a knight and whose name was Zorn.

"What ransom do you demand for the release of Princess Annabelle?" Alexander boldly asked.

"Oh, heavens," laughed the King, "I never had her brought here for *ransom!*"

"You mean you had worse motives?" asked Alexander, in some alarm.

"Oh no, dear boy," replied the King, in a reassuring tone. "Princess Annabelle is indeed as lovely a lady as can be, and she will be the ideal bride for you, if you are clever enough to win her back."

"Then why *did* you have her kidnapped?" Alexander asked.

"The reason will surprise you," replied Zorn. "You have quite a reputation for solving puzzles. I deliberately had the princess brought here to test your skill. You have done well in determining that your princess is on the island, but the difficult part is yet to come."

"And what is that?" asked Alexander.

"Ah!" said the King, "your next task is to find out whether my Grand Vizier is a knight or a knave. If you succeed, I will have Annabelle released. You may ask the Vizier as many questions as you like, but they must all be answerable by yes or no."

"But that's ridiculously easy!" cried Alexander. "I have merely to ask one question—one question whose answer I already know, such as whether two plus two equals four. From his answer, I will of course know whether he is a knight or a knave."

"You shouldn't have interrupted!" said the King. "Of course you can find out by asking just one question whose answer you already know. But I was about to say that you are not allowed to ask any question whose answer you already know."

The suitor stood lost in thought.

"Let me be more explicit," said the King. "You don't have to plan the sequence of questions in advance; at any stage, the question you decide to ask may depend on the answers already given, but at *no* stage are you allowed to ask a question whose truthful answer could be known to you."

The suitor thought about this some more.

"Are you certain that this puzzle can be solved?" he finally asked.

"I never said it was possible," replied the King.

"Oh, come now!" said Alexander in great agitation. "Isn't it unfair of you to give me an impossible task?"

"I never said it was *impossible* either," replied the King. "It's up to *you* to find out whether it's possible. If it is possible, then you must solve the puzzle to win back your princess. If it is impossible, and you can *prove* to me that it is impossible, then again I promise to release Princess Annabelle. Either way you win her back. Those are my terms."

Alexander thought over the problem for many days, then requested an audience with King Zorn.

"I have a strategy," said Alexander. "I need at most two questions!"

"What are the questions?" asked the King.

"Well," replied Alexander, "first I would ask the Vizier if he is a married knight. As of now, I have no means of knowing whether he is married, or whether he is a knight. If he should answer no, then no further questions are necessary; he must be a knight, because a knave is certainly *not* a married knight, hence could not give the truthful answer no to the question."

"But suppose he answers yes?" asked the King.

"In that case, I would know that he is either a married knight or a knave—possibly married, possibly unmarried—because if he is a knight, his answer would be correct, which means he is a married knight; if he is not a knight, then he is a knave. And so a second question would be necessary. I would then ask him, 'Are you an *unmarried* knight?' If he also answers yes to that question, then of course he is a knave (since a knight would never make two incompatible claims). If he answers no, then he must be a married knight (since a knave is not an unmarried knight and hence couldn't give the truthful answer no). And so a no answer would indicate that he is a knight. Therefore, I would know whether he is a knight or a knave."

"Have you actually *tried* this strategy on the Vizier?" asked King Zorn.

"Not yet," replied the suitor. "But it's what I plan to do."

"It's a good thing you didn't," said the King. "Your questions don't fulfill the conditions I have stipulated!"

The King was right. Why does this strategy fail to meet his requirements?

The suitor's strategy might *happen to work if the suitor were lucky, but it is not* bound *to work—for this reason. Suppose Alexander had tried this strategy on the Vizier and had gotten no for the answer to his first question. Then (as Alexander correctly explained) he would know that the Vizier was a knight. However, if the Vizier had answered yes, then although Alexander would certainly know after the second question whether the Vizier was a knight or a knave, the second question would not satisfy the King's requirement that the suitor not know the correct answer before asking the question. The Vizier could answer yes to the first question only if he were either a married knight or a knave (married or not)—in other words only if he were* not *an unmarried knight! And so the truthful answer (no) to the second question would already be determined!*

"Does this mean I will never see my Annabelle again?" asked Alexander mournfully after the King had explained the strategy's inadequacy.

"I didn't say that!" said King Zorn. "You haven't actually *taken* the test yet; you have merely told me what you *would* have done had you taken it. Go back and think about it some more. Then when you are ready, request a formal audience with me and the Grand Vizier, and either determine in my presence, under the terms I have specified, whether he is a knight or a knave, or else prove to my satisfaction that the task is impossible."

Alexander thanked the King, withdrew, and thought the matter over for several more days. He then came to the conclusion that the task was impossible. However, he was afraid to bring his proof before the King because it might contain some subtle error. "If only I had some wise person with whom I could discuss this before taking the test," he thought.

Fortunately for him and Annabelle, Alexander struck up a warm friendship with a world-famous biologist named Professor Bacterius who was visiting the island at the time. Though a scientist by training, Bacterius was an enormously erudite person with a keen interest in logic. The suitor explained his dilemma to Bacterius.

"I believe my proof is correct," Alexander said, "but I would be infinitely grateful to have your opinion."

"With pleasure," replied Bacterius.

"Well," said Alexander, "here is how I see it: I am asking the Vizier questions. Let's consider my last question. This question must be such that a yes answer will indicate that he is of one type and a no answer will indicate that he is of the other."

"Right so far," said Bacterius.

"Furthermore, I would surely know this fact *before* I asked the question," continued the suitor. "Let's suppose that a yes answer will reveal that he is a knight and a no answer will reveal that he is a knave. Since I will know that a knight will answer yes and a knave will answer no, then I will know that yes is the *truthful* answer, and therefore I *will* know the truthful answer to the question before I ask it, which is precisely the thing I am forbidden to do! Therefore, the task cannot be possible."

Bacterius knit his brow and thought for some time before giving his answer.

"I still don't know whether the task is possible," he finally said. "But if I were you, I would *not* give this proof to the King. It has a subtle though definite weakness, and I'm not sure whether it can be remedied."

He then explained the gap in the suitor's proof, and Alexander realized that Bacterius was right.

Can you spot the gap?

There are two fallacies in the proof. In the first place, when the suitor comes to ask the last question, why would he necessarily know that it will be the last question? The question might be of such a nature that if it were answered one way, Alexander could tell whether the Vizier is a knight or a knave, but if it were answered the other way, then further questions would be necessary. However, even if the suitor did know that this was to be the last question, there is a difficulty more serious yet! There is a way that Alexander could know that a yes answer would indicate that the Vizier was of one type and a no answer would indicate that he was of the other. And it might seem very strange, but he could know this only after getting an answer. There is a question having this curious property, as you will soon see!

Alexander was, of course, crestfallen to realize that his proof didn't hold water. Moreover, he was in a quandary as to what to do next; he could not find an airtight argument that the task was impossible, nor could he see any way of accomplishing it. Even Professor Bacterius, with his impeccable logic, was unable to solve the problem!

At this point, fate proved helpful. Alexander, through a combination of craft and bribery, was able to find out where Annabelle was imprisoned and, in the dead of night, sneaked in to visit her. Now, I should tell you that Annabelle, though possessed of little formal education, was widely read, extremely intelligent, and had that most precious mathematical gift of all—a remarkable intuition! When Alexander explained the problem to her, Annabelle solved it instantly.

"Dear boy," she said, "you have nothing to worry about. You can find out in only *one* question whether the Vizier is a knight or a knave, and if you follow my plan, you will not know the truthful answer to the question at the time you ask it!"

She then explained her plan, and Alexander was delighted. The very next day he obtained a formal audience with the King. The entire court was present in full regalia to see how the suitor would fare.

"I am ready," said Alexander. "I claim the task *is* possible and I shall accomplish it by asking the Vizier just *one* question!"

"Now, this I must see!" said the King with a chuckle. (He chuckled because he happened to believe that the task was impossible.) "Proceed!"

The suitor (in accordance with Annabelle's instructions) took a deck of cards out of his pocket, shuffled them thoroughly, took out a card at random and, without looking at its face, showed it to the Vizier. "Is this a red card?" he asked. As soon as the Vizier answered (either yes or no), the suitor looked at the face of the card for the first time and *then* he knew whether the Vizier had lied or told the truth.

"Amazing!" said the King. "I would never have thought of anything like that! But somehow it seems like cheating!"

"Not really," replied Alexander. "When I asked the Vizier whether the card was red, I hadn't looked at it yet, hence I didn't know whether it was red. Therefore, my question was perfectly legitimate according to your terms."

Well, the King had to acknowledge that Alexander had won, and so Annabelle was released.

"Good luck to both of you," said the King. "Why don't you stay another day or so on this island? You haven't yet met our Sorcerer, and he is a most amazing character! He knows about you both and would like to meet you. Why not pay him a visit?"

The happy couple thought this a good idea, and decided to visit the Sorcerer that very afternoon. But that's a story for the next chapter.

4
HOW KAZIR WON
HIS WIFE

ANNABELLE AND ALEXANDER were tired after their long, winding climb to the Sorcerer's castle. But the Sorcerer, whom they discovered to be a most delightful and hospitable chap, served them a delicious brew consisting of equal parts of Ceylon tea and Chinese cinnamon wine, and they were instantly revived.

"Tell me," said Annabelle, whose interests were generally quite practical, "how did you ever establish your reputation here as a Sorcerer?"

"Ah, that's an amusing tale," said the Sorcerer, rubbing his hands. "I started my career here just twelve years ago. But I really owe my beginnings to the philosopher Nelson Goodman, who taught me a clever logical trick some forty years ago."

"What's the trick?" asked Alexander.

"Do you realize," said the Sorcerer, "that despite the fact that every inhabitant of this island is either a knight who only tells the truth or a knave who only tells lies, you can discover the truth or falsity of any proposition simply by asking any inhabitant only one question? And the odd part is that after he answers, you won't know whether his answer is the truth or a lie."

"That does sound clever," said Alexander.

"Well, that's how I got my position here," said the Sorcerer with a laugh. "You see, the first day I set foot on this island, I decided to go to the King's palace and apply for the job of sorcerer. The problem was I didn't know just where the palace was. At one point I came to a fork in the road; I knew that one of the two branches led to the palace, but I didn't know which one. There was a native standing at the fork; surely *he* would know which road was correct, but I didn't know whether he was a knight or a knave. Nevertheless, using the Goodman principle, I was able to find the correct road by asking him only one question answerable by yes or no."

What question did the Sorcerer ask him?

If you ask the native whether the left road is the one that leads to the palace, the question will be useless, since you have no idea whether he is a knight or a knave. The right question to ask is "Are you the type who would claim *that the left road leads to the palace?" After getting an answer, you will have no idea whether he is a liar or a truth-teller, but you* will *know which road to take. More specifically, if he answers yes, you should take the left road; if he answers no, you should take the right road. The proof of this is as follows.*

Suppose he answers yes. If he is a knight, then he has told the truth; he is *the type who would claim that the left road leads to the palace. Hence the left road is the road to take. On the other hand, if he is a knave, his answer is a lie, which means he is* not *the type who would claim that the left road leads to the palace; only one of the opposite type—a knight—would claim that the left road leads to the palace. But since a knight would claim that the left road leads to the palace, then again the left road really does lead to the palace. Regardless of whether the yes answer is the truth or a lie, the left road leads to the palace.*

Suppose now that the native answers no. If he is truthful, then he is really not *the type who would claim that the left road*

*leads to the palace; only a knave would claim that. And since
a knave would make that claim, the claim is false, which means
that the left road does not lead to the palace. On the other hand
if he is lying, then he really would claim that the left road leads
to the palace (since he says he wouldn't), but, being a knave,
his claim would be false, which means that the left road doesn't
lead to the palace. This proves that if he answers no, the right
road is the one to take, regardless of whether the native lied or
told the truth.*

*The Goodman principle is really a remarkable thing. Using
it, one can extract any information from one who either always
lies or always tells the truth. Of course, this strategy won't work
on someone who sometimes tells the truth and sometimes
lies.*

"And so," said the Sorcerer, "I found the correct road and
took it. I got lost a few more times along the way, but even that
was for the best. The native I asked was so astounded at what
I had done that he immediately told many of the island's other
inhabitants, and the news reached the King before I did. The
King was absolutely delighted with the incident and hired me
on the spot! I've been doing good business ever since.

"And now," continued the Sorcerer, as he removed a beauti-
fully bound volume from a shelf, "I have here an exceedingly
rare and curious old book, written in Arabic, known as the
Tellmenow Isitsöornot. I learned of it from the American au-
thor Edgar Allan Poe, in his story entitled 'The Thousand-and-
Second Tale of Scheherazade.' Poe found this odd tale of
Scheherazade in the *Isitsöornot,* but there are many other remarkable tales that he never mentioned and that are still com-
paratively unknown. For example, have you heard of the 'Five
Tales of Kazir'?"

The two guests shook their heads.

"I thought not. These stories are of particular interest to logicians. Here, I will translate them for you.

"Once there was a young man named Kazir whose main ambition in life was to marry a king's daughter. He obtained an audience with the king of his country and frankly confessed his desire.

" 'You seem like a personable young man,' said the King, 'and I am sure my unmarried daughter will like you. But first you must pass a test. I happen to have two daughters named Amelia and Leila; one is married and the other is not. If you can pass the test, and if my unmarried daughter approves of you, then you may marry her.'

" 'Is Amelia or Leila the married one?' asked the suitor.

" 'Ah, that is for you to find out,' replied the King. 'That is your test.'

" 'Let me explain further,' he continued. 'My two daughters are identical twins, yet temperamentally poles apart: Leila always lies and Amelia always tells the truth.'

" 'How extraordinary!' exclaimed the suitor.

" 'Most extraordinary, indeed,' replied the King. 'They have been like this from early childhood. Anyway, when I strike the gong, both daughters will appear. Your task is to determine whether Amelia or Leila is the married one. Of course you will not be told which is Amelia and which is Leila, nor will you be told which of them is married. You are allowed to ask just one question to just one of the two; then you must deduce the name of my married daughter.' "

"Oh, I get it," interrupted Annabelle. "Kazir used the Goodman principle. Is that it?"

"No," replied the Sorcerer. "The suitor happened to know the Goodman principle—though not by that name, of course. His tutor, a venerable old dervish, had taught it to him years before. And so Kazir was overjoyed and thought, 'All I need do is ask either daughter whether she is the type who would claim that Amelia is married. If she answers yes, then Amelia is married; if she answers no, then Leila is married. It's as simple as that.'

"But it *wasn't* as simple as that," continued the Sorcerer. "It so happened that the King could tell from the suitor's triumphant expression that he knew the Goodman principle. So the King said, 'I know what you're thinking, but I'm not going to let you use that logical trick on me. If your question contains more than three words, I'll have you executed on the spot!'

" 'Only three words?' cried Kazir.

" 'Only three words,' repeated the King.

"The gong was struck, and the two daughters appeared."

What three-word question should the suitor ask to determine the name of the King's married daughter?

The suitor should ask, "Are you married?"

Suppose the daughter to whom he puts the question answers yes. She is either Amelia or Leila, but we don't know which. Suppose she is Amelia. Then her answer is truthful; Amelia really is married. But suppose she is Leila. Then her answer is a lie; Leila is not married, so it must be Amelia who is married. Regardless of whether she is Amelia or Leila, if she answers yes, then Amelia must be married.

Suppose, on the other hand, that the daughter answers no. If she is Amelia, then her answer is truthful, which means that Amelia is not married; hence Leila is married. On the other hand, if Leila answers, then her answer is a lie, which means that she, Leila, is married. Regardless of whether the answer no is true, Leila is the married daughter.

"Did the suitor pass the test?" Annabelle asked.

"Alas, no," replied the Sorcerer. "If he hadn't been restricted to a three-word question, he would have had no trouble at all. But, confronted with a totally new situation, he was completely flustered. He just stood there, unable to utter a word."

"So what happened?" asked Alexander.

"He was dismissed from court. But then a curious thing happened. The unmarried daughter had taken a liking to Kazir, and pleaded with her father to summon him back the next day to take another test. Somewhat reluctantly, the King agreed.

"The next day, when the suitor entered the throne room, he exclaimed, 'I've thought of the right question.'

" 'Too late now,' said the King. 'You'll have to take another test. When I strike the gong, again my two daughters will appear (veiled, of course). One will be dressed in blue and the other in green. Your task now is *not* to find out the *name* of my married daughter, but which of the two—the one in blue or the one in green—is *unmarried*. Again you may ask only one question, and the question may not contain more than three words.'

"The gong was struck, and the two daughters appeared."

What three-word question should the suitor ask this time?

The question "Are you married?" will be of no help in this situation; the question to ask is "Is Amelia married?" If the daughter addressed answers yes, then she is married, regardless of whether she lies or tells the truth; if she answers no, then she is not married.

Suppose she answers yes. If she is Amelia, then her answer is truthful; Amelia is married. On the other hand, if she is Leila, then her answer is a lie; Amelia is not married, so Leila is married. Thus, a yes answer indicates that the daughter who answers is married. (I leave it to the reader to verify that a no answer indicates that the person addressed is not married.)

Of course, the question "Is Leila married?" would serve equally well; a yes answer would then indicate that the daughter who answers is not *married, a no answer that she is.*

As the Sorcerer explained to Annabelle and Alexander, there is a pretty symmetry between this problem and the last: to find out if the daughter addressed is married, you ask, "Is Amelia married?"; whereas if you wish to find out whether Amelia is married, you ask, "Are you married?" These two questions have the curious property that asking either one will enable you to deduce the correct answer to the other.

"Again the suitor failed," continued the Sorcerer, "but the unmarried daughter was more fond of him than ever. The King could not resist her pleadings, and so he agreed to give Kazir a third test the next day.

" 'This time,' said the King to the suitor, 'when I strike the gong, just one daughter will appear. Your task now is to find out her first name. Again you may ask only one question, and it may contain no more than three words.' "

What question should the suitor ask?

This problem is simpler than the preceding two. All the suitor need ask is any three-word question whose answer he already knows, such as "Does Leila lie?" To this particular question, Amelia would obviously answer yes and Leila would answer no.

" 'Now really,' said the King to his daughter after the suitor failed the third test, 'are you sure you want to marry him? He strikes me as quite a simpleton. The last test was ridiculously easy, and you know it.'

" 'He was just nervous,' responded the daughter. 'Please try him once more.'

"Well," said the Sorcerer, "for the next test the suitor was told that after the gong was struck the unmarried daughter would appear. The suitor was to ask her just one three-word question answerable by yes or no. If she answered yes, the suitor could marry her. If she answered no, he could not."

What question should the suitor ask?

The question "Are you Amelia?" would work perfectly well. Amelia, who is truthful, would answer yes, and Leila, who lies, would also answer yes—that is, she would falsely claim to be Amelia.

" 'This is trying my patience,' said the King to his daughter after the suitor had failed a fourth time. 'No more tests!'

" 'Just one more,' begged the daughter. 'I promise it will be the last.'

" 'All right, the very last one, you understand?'

"The daughter promised not to ask for any more tests, and so the King agreed.

" 'Now then,' said the King quite sternly to the suitor (who was trembling like a leaf), 'you have already failed four tests. You seem to have difficulty with these three-word questions, and so I will relax that requirement.'

" 'What a relief!' thought the suitor.

" 'When I sound the gong,' said the King, 'again only one daughter will appear; she may be the married one or the unmarried one. You are to ask her only one question answerable by yes or no, but the question may contain as many words as you like. From her answer, you must deduce *both* her name *and* whether she is married.' "

Can you think of a question that will work?

The king (evidently sick by now of the whole business) has given the suitor an impossible task. In each of the first four tests, the suitor had to determine which of two possibilities held. In this test, however, he is to determine which of four possibilities holds. (The four possibilities are: 1. the daughter

*addressed is Amelia and married; 2. she is Amelia and unmar-
ried; 3. she is Leila and married; 4. she is Leila and unmarried.) However, there are only two possible responses to the
suitor's question (either yes or no, since the question is required to have a yes/no answer). And with only two possible
responses, it is impossible to determine which of four possibilities holds.*

*When I say that the task is impossible, I merely mean that
there is no yes/no question that is bound to work. There are
several possible questions (at least four) that might happen to
work if the suitor is lucky. For example, consider the question "Is
it the case that you are married and that your name is Amelia?"
Leila would answer yes to this (regardless of whether she is
married because a compound statement—one with two parts
joined by "and"—is false if both parts or only one part of the
statement is false); Amelia, if she is married, would answer yes
and, if she is unmarried, would answer no. So a yes answer would
leave the suitor in the dark, whereas a no answer would indicate
for sure that the one addressed is Amelia and unmarried. So if
the suitor asked that question, he would have a twenty-five
percent chance of finding out which one of the four possibilities
actually holds. But no single question could ensure a certainty of
finding out which one of the four possibilities holds.*

"So they never did get married?" asked Annabelle.

"The King never consented," replied the Sorcerer. "But the
daughter was so furious at the unfairness of the last question
that she felt perfectly justified in marrying Kazir without her
father's permission. The two eloped and, according to the *Isitsöornot*, lived happily ever after.

"There is a valuable lesson to be learned from all this,"
continued the Sorcerer. "Never rely too much on general principles and routine mechanical methods. Certain classes of
problems can be solved by such methods, but they completely

lose their interest once the general principle is discovered. It is, of course, good to know these general principles—indeed, science and mathematics couldn't advance without them. But to depend on principles while neglecting intuition is a shame. The King was very clever to construct his tests so that the mere knowledge of a general principle—in this case the Goodman principle—would be of no avail. Each of his tests required a bit of ingenuity. Here the suitor failed, because of a lack of creative thinking."

"I would like to know one more thing," said Annabelle. "Is it recorded which of the two daughters Kazir married?"

"Oh, yes," replied the Sorcerer. "Fortunately for the suitor it was Amelia, the truthful one. This, I believe, contributed to their living happily ever after.

"The story of Leila's marriage," he continued, "was also recorded in the *Isitsöornot*, and I find this story particularly droll. It seems that Leila detested her suitor, but when he asked her one day, 'Would you like to marry me?', she, being a perpetual liar, said yes. And so they were married.

"So you see," concluded the Sorcerer, "perpetual lying sometimes has its dangers!"

Well, they all had a hearty laugh over that story. Then Annabelle and Alexander rose and thanked the Sorcerer for a most entertaining and instructive afternoon, explaining that they had to get back and make preparations for their departure the next day.

"Have you heard about the epidemic that hit this island three years ago?" asked the Sorcerer.

They shook their heads.

"Oh, I would like to tell you all about it, but unfortunately I can't today. I'm expecting a visit from the island Astrologer any minute. Why don't you stay a few more days?"

"My parents will be worried about me," explained Annabelle.

"Not so," replied the Sorcerer. "I've already dispatched a message to your island explaining that you are here and well. Your father knows that I am a knight."

This reassured the couple, and they agreed to visit the Sorcerer the next day.

"Who is this Astrologer?" asked Alexander, as they were leaving.

"Oh, he's a complete imbecile, as well as a knave and a charlatan. I have to put up with him; it's a matter of island politics. But I advise you to have nothing to do with him."

Just then, the Astrologer entered the room, nodded to the couple (whom he had never seen before), and said to the Sorcerer, "She is the Queen of Sheba and he is King Solomon, you know."

"Typical," said the Sorcerer to the departing couple with a knowing wink.

5

A PLAGUE OF LIES

"IT CAME LIKE A SIROCCO WIND and struck about half of the island's inhabitants. Fortunately, I was among those who were spared," the Sorcerer explained to Annabelle and her suitor, Alexander. They had both made a special trip to the Sorcerer's tower to hear his account of the strange epidemic that had swept across the Island of Knights and Knaves about three years before.

"Nobody could really diagnose it," continued the Sorcerer. "Even Professor Bacterius, with all his knowledge of immunology, was totally baffled."

"Was it viral or bacterial?" asked Annabelle.

"Even that was unknown!" exclaimed the Sorcerer. "There was absolutely no observable change in the body chemistry. In fact, there were no physical symptoms at all; the effects were purely psychological. The whole epidemic lasted only a week, during which time pandemonium reigned over the island. Then, quite suddenly, things returned to normal."

"You say the symptoms were purely psychological," said Alexander. "Just what were they?"

"Well," replied the Sorcerer, "all those who were struck reversed their normal lying or truth-telling role. No longer was it true that all knights told the truth and that all knaves lied.

Instead, sick knights lied and sick knaves told the truth, while healthy knights continued to tell the truth and healthy knaves continued to lie. So during this strange week, there were *four* types on the island—healthy knights, sick knights, healthy knaves, and sick knaves. Healthy knights and sick knaves told the truth; sick knights and healthy knaves lied. So if you came across a native who made a statement you knew to be false, in general you could not know whether he was a sick knight or a healthy knave."

"That must have been frightfully confusing," said Annabelle.

"At first, yes," replied the Sorcerer. "But then I went around the island interviewing natives I didn't know to see what I could learn, and I learned some interesting things indeed."

Alexander thought for a bit, and then commented, "I can see that *in general* you couldn't tell a sick knight from a healthy knave. For example, if a person said that two plus two equals five, you could certainly not tell whether he was a sick knight or a healthy knave. But were there exceptions? Was it *sometimes* possible to tell from just one false statement whether the speaker was a sick knight or a healthy knave?"

"Good question," said the Sorcerer. "I actually encountered such a situation. I came across a native I didn't know. He made a statement from which I could deduce not only that it was false, but even whether he was a sick knight or a healthy knave."

Can you think of one such statement?

One statement that would work is "I am a sick knight." The statement can't be true, since a sick knight wouldn't truthfully claim to be a sick knight. Therefore, he is lying, which means he is either a sick knight or a healthy knave (since these are the only types who lie). But he is not a sick knight. Hence, he is a healthy knave. Of course, the statement "I am a healthy

*knave" would work equally well; only a sick knight could make
that statement.*

"On another occasion," continued the Sorcerer, "I came
across a native who made a statement from which I could
deduce that he must be a knave, though I couldn't tell whether
he was sick or healthy."

Can you supply such a statement?

*A simple statement that works is "I am sick." A healthy knight
would never lie and claim to be sick, and a sick knight would
never truthfully admit to being sick. Hence, a native who
would say that must be a knave. He could be either a healthy
knave who lied about being sick or a sick knave who truthfully
claimed to be sick.*

"Soon after, I met another native who made a statement
from which I could deduce that he must be sick, but not
whether he was a knight or a knave," said the Sorcerer.

What statement would work?

*The statement "I am a knave" would work. (I leave the proof
to the reader.)*

"I next came across a native who made a statement from
which I could deduce that he was either a healthy knight, a sick
knight, or a healthy knave, but it was impossible to know which
of the three cases held."

What statement would work?

*The statement "I am a healthy knight." A healthy knight could
truthfully say that; a sick knight or a healthy knave could lie
and say that; but a sick knave could never lie and say that he*

is a healthy knight. Thus, anyone but a sick knave could make that statement.

"Finally, I came across two natives unknown to me, named Astor and Benedict. First Astor said something I could not understand, since he used the island's native tongue, which I had not fully mastered. Then I asked Benedict what Astor had said. Benedict replied, 'He said that he is either a sick knight or a healthy knave.' Astor protested, 'I never said that!' But Benedict said, 'Astor is a knave.' And finally Astor said, 'Benedict is sick!'

"This enabled me to completely classify Astor and Benedict," said the Sorcerer.

What are they?

During this strange week, no inhabitant could possibly claim to be a sick knight or a healthy knave, because if he is a sick knight or a healthy knave, he lies; hence, he would never truthfully admit to being a sick knight or a healthy knave. And if he were not a sick knight or a healthy knave, then he would be truthful; hence, he would never falsely claim to be a sick knight or a healthy knave. Therefore, Benedict lied when he said that Astor claimed to be a sick knight or a healthy knave. Thus, Astor must be truthful, because he correctly denied having said that. Now, Benedict, who lies, said that Astor is a knave, so Astor is really a knight. Since Astor is a knight and is truthful, Astor must be a healthy knight. Also, Astor, who is truthful, said that Benedict is sick. Therefore, Benedict really is sick. Since Benedict lies and is sick, he must be a sick knight. So the answer is that Astor is a healthy knight and Benedict is a sick knight.

"It so happens," continued the Sorcerer, "that the island Astrologer was attacked by the disease. As you know, he is a knave, and usually a rather stupid one. But the disease had a miraculous effect on him. Not only did he tell the truth all week, but he was eminently sensible during the entire period. It's really a pity he recovered.

"Anyway, during this time, we did a good deal of research together. On one particular day we were walking on the beach when we spied a native walking in our direction. 'Oh, I know him,' said the Astrologer. 'I can tell you whether he is a knight or a knave, but not whether he is sick.'

"As the native passed us, the native said, 'I am a healthy knight.' The Astrologer then said to me, 'Good. I now know whether he is sick.' "

The Sorcerer asked Annabelle and Alexander, "Was the native a knight or a knave, and was he sick or healthy?"

"Just a minute," said Annabelle. "Are you sure you gave us enough information?"

"I certainly am," replied the Sorcerer.

What is the solution?

. . .

From the fact that the native claimed to be a healthy knight, all that follows is that he is not a sick knave. (We saw in an earlier problem that a healthy knight, a sick knight, or a healthy knave could all make that statement, but that a sick knave could not.) Now, before the Astrologer heard the native talk, he knew whether the native was a knight or a knave, but we are not told which. Suppose he previously knew that the native was a knight. Then after the native made his statement, the Astrologer would have no way of knowing whether the native was a sick knight or a healthy knight: he would have known no more than he knew before. But since we are told that the Astrologer did finally know, then the only possibility is that he previously knew the native to be a knave and finally knew him to be a healthy knave.

. . .

"More interesting yet," continued the Sorcerer, "while on our walk, the Astrologer and I came across another native talking to himself, about whom we each had partial information. 'I know whether he is a knight or a knave,' I told the Astrologer, 'but I don't know whether he is sick.'

" 'That's curious,' replied the Astrologer. 'I happen to know whether he is sick, but I don't know whether he is a knight or a knave. Don't tell me what you know, and I won't tell you what I know. Instead, let's go over and hear what he has to say.'

"So we went over to the native, only to hear him mumbling to himself, 'I am not a healthy knave.'

"The Astrologer and I both thought for a while, but I could not yet tell whether he was sick or healthy, nor could the Astrologer yet tell whether he was a knight or a knave. Was the native a knight or a knave, and was he healthy or sick?"

"Now, wait a minute," said Annabelle. "If you don't know, then how do you expect us to know?"

"I didn't say I don't know," replied the Sorcerer. "I said I *didn't* know at the time. It was only when the Astrologer later told me that *he* still didn't know that I knew."

What is the solution?

A healthy knave could lie and claim he is not a healthy knave; a sick knave could truthfully claim he is not a healthy knave; a healthy knight could truthfully claim he is not a healthy knave; but a sick knight could never make the true statement that he is not a healthy knave. So, what follows from the native's statement is simply that he is not a sick knight. After the native spoke, then, both the Sorcerer and the Astrologer knew that he was not a sick knight.

Now, the Sorcerer had previous knowledge of whether the native was a knight or a knave. If he had previously known him to be a knight, then after knowing that he was not a sick knight, he would have known him to be a healthy knight. But we are told that the Sorcerer could not tell from the native's statement whether he was sick or healthy; therefore, the Sorcerer must have previously known the native to be a knave.

The Astrologer, on the other hand, previously knew whether the native was healthy or sick. Had he previously known him to be sick, then after the native spoke, he would have known that he was a sick knave (since he knew that the native was not a sick knight), but if he had previously known him to be healthy, he would have still been in the dark after the native spoke. Since he was still in the dark, then it must be that he previously knew that the native was healthy. Therefore, the native was a healthy knave.

"During this epidemic," said the Sorcerer, "a valuable ring of mine was stolen. Three suspects, Jacob, Karl, and Louie, were immediately arrested, and there was a trial that very day. Everybody knew that one of the three was guilty, but before the trial the court did not know which of the three it was. Here is a transcript of the trial. (I should mention that the three defendants were close friends, and it was correctly assumed that the two innocent ones knew who the guilty one was.)"

Judge (to Jacob): What do you know about the theft?
Jacob: The thief is a knave.
Judge: Is he healthy or sick?
Jacob: He is healthy.
Judge (to Karl): What do you know about Jacob?
Karl: Jacob is a knave.
Judge: Healthy or sick?
Karl: Jacob is sick.

"The judge thought for a while and then asked Louie, 'Are you by any chance the thief?' Louie answered (yes or no), and the judge then decided the case. Who stole the ring?"

"Just a minute," said Alexander. "You haven't told us what Louie answered."

"You can solve the problem without being told," replied the Sorcerer.

Who stole the ring?

Karl's two answers were either both truthful or both lies. If they were both truthful, then Jacob is a sick knave; if they were both lies, then Jacob is a healthy knight. So Jacob is either a sick knave or a healthy knight, and, in either case, he is truthful in his present condition. Hence, Jacob's answers were both truthful, so the thief is actually a healthy knave and hence a liar. Since Jacob is truthful and the thief is a liar, Jacob is not the thief.

The judge knew this even before she interrogated Louie. Then she asked Louie whether he was guilty, but we are not told what Louie answered. If Louie answered yes, then he must be innocent, for we already know that the real thief would lie about being guilty. On the other hand, if Louie answered no, then there is no way of telling whether he is innocent or guilty. If he answered truthfully, he would be innocent, which is perfectly consistent with the fact that the thief lies. But if he answered falsely, then he is the thief, which is again consistent with the thief's being a liar. So if Louie answered no, the judge couldn't have made a conviction. But the judge did make a conviction. Thus, Louie must have answered yes, proving to the judge that Louie was innocent. So Karl stole the ring.

By a curious turn of fate, Louie acquitted himself by claiming to be guilty!

6

ON THE OTHER HAND

THE WEDDING of Princess Annabelle and her suitor, Alexander, on the Island of Knights and Knaves went smoothly until the time came for the couple to take their vows. Then the justice of the peace declared: "I do *not* now pronounce you man and wife." Everyone gasped in horror. Even Annabelle and Alexander—who were used to the odd customs of the Island of Knights and Knaves, where every inhabitant is either a knight, who always tells the truth, or a knave, who always lies—held their breath. Then, at last, among the fainting guests, a knight in the wedding party assured everyone that the justice was in fact a *legally certified* knave and that his statement therefore constituted legal evidence that the couple was married. A sigh of relief went up. And then a chorus of cheers.

King Zorn, who had grown quite fond of the couple, provided a magnificent banquet at the palace. Everyone of importance attended, including, of course, the Sorcerer, and even the doddering, grumpy Astrologer. After the feast came the dancers, jugglers, fire-eaters, and magicians. But the evening's real entertainment began when the Sorcerer gathered everyone about him to hear of his travels.

"I visited an odd country some months ago," he began,

"where every inhabitant is either right-handed or left-handed."

"What's so odd about that?" interrupted the Astrologer.

"Nothing," replied the Sorcerer, "except that whatever right-handed inhabitants write with the right hand is true, but whatever they write with the left hand is false. Left-handed inhabitants are the opposite: whatever they write with the left hand is true, and whatever they write with the right hand is false."

"That's not strange," said the Astrologer. (The Astrologer, remember, is a knave.)

"It certainly *is* strange," said the King quite sharply, "and I wish you would stop interrupting!" There was a tense silence.

"Go on with your story," said the King. "The Astrologer won't interrupt again, or I'll have him thrown into the dungeon!"

"Well," said the Sorcerer, "this country provided marvelous opportunities for logical detective work. The problems that arose were just the sort that interest Your Majesty."

"Such as?" asked the King.

"On my very first day there, I came across a scrap of paper on which an inhabitant had written a single sentence. From that sentence I could deduce that the inhabitant must be left-handed."

"What was the sentence?" asked the King.

"Ah! That's the question I would like you all to answer," replied the Sorcerer.

A sentence that works is "I wrote this with my left hand." The sentence is either true or false. If it is true, then the writer really did write it with his left hand. And since only left-handed inhabitants write true sentences with the left hand, he must be left-handed. But if the sentence is false, then the writer wrote it with his right hand. Since only left-handed inhabitants write false sentences with the right hand, again the writer must be left-handed.

"The next day," said the Sorcerer, "I came across a sentence jotted on a slip of paper from which I deduced that the writer was right-handed and had written the sentence with his left hand. What sentence did he write?"

"I know!" said one of the guests. "The sentence was something like 'Two plus two is five.' That could have been written by a right-handed person using his left hand."

"Of course it *could* have been," said the Sorcerer, "but it also could have been written by a left-handed person using his right hand. The sentence I came across could have been written *only* by a right-handed person using his left hand."

What was the sentence?

The simplest solution is the sentence "I am left-handed, and I wrote this sentence with my right hand." The sentence must be false, because if it were true, that would mean that a left-handed person wrote a true sentence with his right hand, which is impossible. Since the sentence is false, and it was not written by a left-handed inhabitant with his right hand, it must have been written by a right-handed inhabitant with his left hand.

"After I had spent a week in this country," boasted the Sorcerer, "I gained quite a reputation as a handwriting expert. One day the Chief of Police consulted me on a case. Two scraps of paper had been found. On one was written: 'I always write with my right hand.' On the other was written: 'I sometimes write with my left hand.'

"The police knew who the writer was, and for some reason, about which it was not my business to inquire, it was very important to find out which of the two sentences was true. Was there any way of determining whether the writer always wrote with his right hand or whether he sometimes wrote with his left?"

The two sentences contradict each other; therefore, one of them is true and the other false. That means that the writer sometimes uses one hand and sometimes the other. And so the second sentence is true: he sometimes writes with his left hand.

"On another occasion," said the Sorcerer, "the police consulted me about a robbery. It was known that the thief was left-handed. A suspect had been arrested, and the first thing to be done was to determine whether he was right-handed or left-handed. The police had searched his house and found a notebook with entries that the suspect admitted having written. I have here facsimiles of the first two pages."

The Sorcerer held up the first page, which read: "The sentence on page 2 is false." And then he held up the second, which read: "The sentence on page 1 was written with my left hand."

"The police wanted to know whether this notebook was at all relevant to the case," explained the Sorcerer, passing around the pages. "Was it?"

There is no way of determining which of the sentences is true. But it can be determined whether the suspect is right-handed or left-handed. Suppose the sentence on page 1 is true. Then the sentence on page 2 is false, which means that the sentence on page 1 was written with the right hand. The writer must therefore be right-handed, since he wrote the true sentence on page 1 with his right hand. Now, suppose the sentence on page 1 is false. Then the sentence on page 2 must be true (because the sentence on page 1 falsely says that the sentence on page 2 is false), which means that the sentence on page 1 was written with the left hand. Again, the suspect must be right-handed, since he wrote the false sentence on page 1 with his left hand. Thus, the suspect is right-handed and therefore innocent of the robbery.

"The evening is getting on," said the Sorcerer. "So I will tell you only one more story.

"One afternoon I was sitting at a café with a friend of mine named Robert Smith. I chanced to ask him whether he was right-handed or left-handed. Instead of answering, he smiled, tore a sheet from his notebook, and wrote: 'I am left-handed, and I wrote this with my left hand.' Of course I saw which hand he used, but I was not able to deduce whether he was right-handed or left-handed. Just then his wife joined us. He showed her the sentence he had written and asked whether she could tell which hand he had written it with. She knew whether her husband was right-handed or left-handed, and she was a good logician. Still, she couldn't tell which hand he had used. Is he right-handed or left-handed?"

"Now just a minute," said the King. "You said that you saw which hand Mr. Smith used. Have you forgotten to tell us which hand it was?"

"I didn't forget," said the Sorcerer. "From the information I have given you, it is possible to deduce which hand he used."

Which hand did Mr. Smith use, and is he right-handed or left-handed?

If Mr. Smith were right-handed, he couldn't possibly have written the sentence with his right hand. Therefore, there are only three possibilities: he is right-handed and falsely wrote the sentence with his left hand; he is left-handed and falsely wrote the sentence with his right hand; or he is left-handed and truthfully wrote the sentence with his left hand.

Now, the Sorcerer saw which hand Mr. Smith used. If it

were his right hand, then the Sorcerer would have known that
only the first possibility could be true and that Mr. Smith was
right-handed. However, the Sorcerer didn't know this. Hence,
Mr. Smith must have used his left hand. (The Sorcerer then
couldn't tell whether the first or the third possibility was the
truth.)

Mrs. Smith, on the other hand, knew whether Mr. Smith was
right-handed or left-handed, but she didn't see which hand he
had used to write the sentence. If Mr. Smith were right-
handed, his wife would know this, and she would have been
able to deduce that only the first option was the actual case.
That is, she would have figured that he had written the sen-
tence with his left hand. Since she was unable to do this, her
husband must be left-handed. Even though she knew he was
left-handed, she couldn't tell whether the second or the third
possibility was the actual case. This proves that Mr. Smith is
left-handed and that he wrote the sentence with his left hand.

"And now a toast to the couple!" said King Zorn. "May their
years be long and happy, and may they pay many more visits
to this island, bringing new and challenging puzzles with
them!"

Another round of cheers thundered.

"And now for the main event of the evening—the possible
wedding gift to the bride and bridegroom."

"*Possible* gift?" muttered the Astrologer.

"I shall let the Sorcerer make the presentation," said the
King. "I believe you will all find it amusing."

The Sorcerer rose and walked over to two magnificent treasures that the company had been eyeing all evening.

"The bridegroom is to rise and make a statement," said the
Sorcerer. "His Majesty has graciously promised that if the
statement is true, then he will give the couple one of these two
treasures and possibly both. But if the statement is false, then

His Majesty flatly refuses to give the couple any wedding gift at all!"

The Sorcerer then gave Alexander a knowing wink, as if to say: "Come on now, old boy, use your head and come up with a statement such that the king's only option is to give you *both* treasures!"

Alexander cleverly made such a statement. What did he say?

Alexander said: "It is not the case that His Majesty will give us one and only one of the treasures."

Suppose the statement is false. Then it is the case that the King will give one and only one of the treasures. But the King can't give one of the treasures for a false statement. Therefore, the statement can't be false; it must be true. Since the statement is true, then the King, to keep his promise, must give at least one of the treasures. Since Alexander's statement is true, though, the King can't give only one of the treasures. (If he did, this would falsify the statement!) Hence, the King must give both of the treasures. And so he did.

7

THE ISLAND OF
PARTIAL SILENCE

A FEW DAYS after the wedding our happy couple, Princess
Annabelle and Alexander, set sail for home. But alas, they
were in for trouble! They had only been at sea a few hours
when their vessel was attacked by pirates; the two were cap-
tured and sold as slaves to the king of another island—a grim
place known as the Island of Partial Silence.

"Let me explain this island to you," said the grim King to the
two captives. "As on the island of King Zorn, every inhabitant
here is either a knight or a knave. However, on this island,
people don't always answer questions asked of them! If you ask
a knight a question, if he answers at all, his answer will be
truthful, but he may refuse to answer. If you ask a question of
a knave and he answers at all, his answer will be a lie, but again
he may simply refuse to answer. If you cook up a special
question that a native cannot answer without violating his
knighthood or knavehood, as the case may be, then he will
certainly refuse to answer. You see now why this island has its
name.

"Two plus two equals four," continued the King. "Now you
see that I am a knight. Unfortunately for you, I am not as
pleasant a soul as King Zorn. On the other hand, I am not quite

54

as evil as I am cracked up to be, and so I will give you two a sporting chance. I will give you five extremely difficult tests. If you pass *all* of them, I will set you free; if you fail so much as *one* of them, the gentleman's head will be forfeit and the Princess will be mine!"

"Five extremely difficult tests and we must pass *all* of them, and you call that a *sporting* chance?" asked Alexander.

"Those are my terms!" said the King sternly.

Here are the five tests.

Test 1 • Design a question to which a knave could answer either yes or no, but which a knight cannot answer at all.

Test 2 • Design a question to which a knight could answer either yes or no, but which a knave cannot answer at all.

Test 3 • Design a question to which a knight could answer no but not yes, and which a knave could not answer at all.

Test 4 • Design a question to which a knight could answer yes but not no, and which a knave could not answer at all.

Test 5 • Design a question that neither a knight nor a knave can answer at all.

Solutions to First Five Tests

Test 1. A question that works is: "Is no your answer to this question?" (The question doesn't mean: "Is no the *correct* answer to this question?" It means: "Is no the answer you will actually give to this question?" Perhaps a more precise way of phrasing it is: "After you have answered this question, will the answer have been no?")

As the Sorcerer explained to Abercrombie in Chapter 1, no

one who is asked this question can answer it correctly, hence a knight cannot answer it. A knave can answer either yes or no, since both are wrong answers.

Test 2. The question is: "Is yes your answer to this question?" A yes answer and a no answer are both correct, hence only a knight could answer the question, and he can answer either way.

Test 3. I doubt that any *simple* question will work; the question (I believe) must be compound. A question that works is: "Is it the case that you are a knave *and* that yes is your answer to this question?"

If the one addressed is a knight, then it certainly is *not* the case that he is a knave and that yes is his answer, because it is not even the case that he is a knave! So if he is a knight, the correct answer to the question is no; hence he, being a knight, will answer no. So a knight can answer no, but he cannot answer yes.

Now, suppose the one addressed is a knave. Suppose that he answers yes. Then it is true that he is a knave *and* that yes is his answer, hence the *correct* answer to the question is yes, which means that the knave answers correctly, which a knave cannot do! Therefore a knave cannot answer yes to this question. Suppose the knave answers no. Then it is false that he is a knave *and* that he answered yes (because it is false that he has answered yes); hence his no answer was correct, which is again impossible for a knave! Therefore a knave cannot answer either yes or no to this question.

Test 4. A question that works is: "Is it the case that either you are a knight or that yes is your answer to this question?"

Note: Any statement of the form: "Either this or that" is to be counted true if the "this" and the "that" are both true. For example, if a college entrance requirement states that an entering freshman must have had either a year of mathematics or a year of foreign language, the college certainly won't exclude an applicant who has had both! And so as used throughout this book, either-or will mean *at least one* (and possibly both).

Now let's see why the question works: Suppose the one questioned is a knight. Then it is certainly true that *either* he is a knight *or* his answer is yes. Hence yes is the correct answer to the question, and so the knight can answer yes (but he can't answer no!). Suppose, on the other hand, you are dealing with a knave. If he answers yes, then it is true that *either* he is a knight *or* yes is his answer (because yes is his answer), hence the knave answered correctly, which is impossible. Suppose the knave answers no. Then it is neither the case that he is a knight nor that he answered yes, and so the correct answer to the question is no, which means that the knave answered correctly! So we see that a knave could not answer either yes or no.

Test 5.   A question that works is: "Is it the case that you are either a knight who will answer no to this question or a knave who will answer yes?"

Suppose he is a knight. If he answers no then he is a knight who has answered no; hence yes is the correct answer to the question and the knight has given the wrong answer, which cannot be. If the knight answers yes, then he is neither a knight who has answered no nor a knave who has answered yes (since he is not a knave at all), and therefore no is the *correct* answer to the question. It cannot be that the knight has again given a false answer, so a knight cannot answer the question at all.

As for a knave, suppose he answers yes. Then he is a knave who has answered yes, so that means he is *either* a knight who has answered no *or* a knave who has answered yes (he is in fact the latter). This means that yes is the correct answer to the question, and therefore a knave has answered correctly, which is not possible. Suppose he is a knave who answers no. Then he is neither a knight who has answered no nor a knave who has answered yes; hence the correct answer to the question is no, so again the knave has answered correctly, which he cannot do. And so a knave can't answer the question either.

Note: An alternative question that does the job is: "Are you the type who could claim that no is your answer to this question?"

Those readers who recall the Nelson Goodman principle, explained in Chapter 4, can quickly see why this question works. (The Goodman principle is that whenever you ask a knight or a knave whether he is the type who could claim such-and-such, if he answers yes, the such-and-such must be true; if he answers no, the such-and-such must be false.) Suppose this principle is applied to the above question (where now the such-and-such is "No is your answer to this question"). If he answers yes, then no is his answer, which is absurd. If he answers no then yes is his answer, which is absurd. Therefore he must remain silent.

Alexander was allowed to consult Annabelle on these problems, and together they worked out all five of them correctly. The next day they brought their answers to the King.

"Hm!" he said, after he had checked the correctness of their solutions. "I'm not sure I should let you go yet! Tomorrow I have eleven more tests for you."

Well, it so happened that the King died that night. Some historians have claimed that he died of a heart attack; others, that he died of a bad conscience for playing foul. A third group maintains that he died of a heart attack *caused* by a bad conscience. This is not an easy matter to settle! At any rate, the new king who ascended the throne the next day was a fairminded individual who insisted on honoring the old king's promise, and so the couple were released. They happily set sail once again for home, arrived without further mishap, and were joyfully welcomed by the people of their island.

Now, there are several historians who have wondered just what the eleven tests were that the deceased King had in mind! Of course the matter can never be settled (unless there is an afterlife—as a logician I must think of *all* possibilities!), but I do have a theory that I would like to tell you.

In reading the chronicles where I found this story, I thought it rather odd that the King gave just five tests to the couple instead of sixteen; there are eleven other possibilities in a similar vein that he left out, and my theory is that these are the very ones he was planning to spring the next day. You see, there are four possibilities for what a knight could do with the question (he could answer yes but not no, or he could answer no but not yes, or he could answer either yes or no, or he could answer neither yes or no), and with each of these possibilities there are four possibilities for what a knave can do. Thus there are sixteen possibilities altogether. The interesting thing is that each of these sixteen possibilities can be realized by a question! Five of them have already been considered, and I have checked that for each of the eleven requirements listed below (I have numbered them 6 through 16), there is indeed a question satisfying that requirement.

6. A knave could answer yes but not no, and a knight could not answer at all.
7. A knave could answer no but not yes, and a knight could not answer at all.
8. A knave could answer yes but not no, and a knight could answer either way.
9. A knave could answer no but not yes, and a knight could answer either way.
10. A knight could answer yes but not no, and a knave could answer either way.

11. A knight could answer no but not yes, and a knave could answer either way.
12. Both knights and knaves can answer either way.
13. Both types can answer no; neither type can answer yes.
14. Both types can answer yes; neither type can answer no.
15. A knight can answer no but not yes; a knave can answer yes but not no.
16. A knight can answer yes but not no; a knave can answer no but not yes.

The reader might have fun trying these as exercises. I give answers at the end of this chapter (but I do not give proofs that these answers work; by now the reader should be able to supply these proofs).

DISCUSSION

Twelve of the questions involved (including the first five, which the reader already knows) have the curious property that their *correct* answer cannot be determined until the *actual* answer is already given. (I can't see how questions without this property could possibly work!) They also have the property of *self-reference*—each question is a question about the very question itself!

Another example of a self-referential question is: "Does this question contain exactly seven words?" The answer is obviously yes. Curiously enough, yes is *not* the answer to the question "Does this question contain seven words?" Another example of a self-referential question is: "Is this question a silly question?" (My guess is that the correct answer is yes.)

There is a story told that the Greek philosopher Epimenides once made a pilgrimage to meet the Buddha. Epimenides asked him: "What is the best question that can be asked, and what is the best answer that can be given?" The Buddha replied: "The best question that can be asked is the question you have just asked, and the best answer that can be given is the answer I am giving."

Self-reference plays a prominent role in modern logical theory and computer science. It is also addictively fascinating! It is central in the proof of Gödel's famous incompleteness theorem. (The Sorcerer will have much to say about this later in the book.)

Answers to Exercises 6 through 16

6. Is it the case that you are a knight and that no is your answer to this question?

7. Is it the case that either you are a knave or that no is your answer to this question?

8. Is it the case that you are a knight and that you will answer yes?

9. Is it the case that either you are a knave or you will answer yes?

10. Is it the case that either you are a knight or you will answer no?

11. Is it the case that you are a knave and that you will answer no?

12. Will you answer yes?

13. Are you a knave?

14. Are you a knight?

15. Does two plus two equal five?

16. Does two plus two equal four?

PART II

PUZZLES AND METAPUZZLES

8

MEMORIES OF THE SORCERER'S UNCLE

A few months after arriving home, our couple felt the urge to visit King Zorn's island again—particularly to see the Sorcerer, of whom they had become quite fond. They planned to stay a week or two, little realizing that things would become so interesting that they would be unable to tear themselves away for several months!

Once arrived on the island, they lost no time in calling on the Sorcerer, who was delighted to see them. He was in a happily nostalgic mood, and spent the afternoon reminiscing about his uncle.

"You really should have met him," he said to Annabelle and Alexander. "He was one of the most interesting people I have ever known. It was he, in fact, who got me interested in logic."

"How did he do that?" asked Alexander.

"Quite naturally," replied the Sorcerer. "He gave me all sorts of fascinating puzzles throughout my boyhood and adolescence. I recall when I was extremely young—about six—my uncle had four dogs, with whom I used to play a great deal. One day he gave me a puzzle about them, and I don't know whether the story was true or made up—but this is the puzzle:

· 1 ·

" 'I once set out a bowl of biscuits for my dogs. First, the oldest one came by and ate half of them and one more. Then, the second dog came by and ate half of what he found and one more. Then, the third one came by and ate half of what she found and one more. Then, the fourth and littlest one came by and ate half of what he found and one more, and the biscuits were then all gone. How many biscuits were in the bowl to begin with?'

"That is the problem my uncle gave me."

What is the answer? (From this point on, solutions are generally given at the end of each chapter.)

· 2 ·

"I remember," the Sorcerer continued, "once my uncle asked me: 'Which is more, six dozen dozen or a half a dozen dozen?' "

"That's obvious," said Annabelle.

"Of course it is," said Alexander.

"True," said the Sorcerer, "but many people nevertheless get it wrong."

What is the correct answer?

· 3 ·

"We lived near a farm," said the Sorcerer. "The farmer sold most of his produce to wholesalers, and some he sold retail at a little vegetable stand. My uncle told me that the farmer sold 90 percent of his produce wholesale and 10 percent retail, but he got twice as much per item retail as when he sold the item wholesale. My uncle then asked me if I could figure out what percentage, or what fraction, of his gross income came from the stand."

What is the answer?

• 4 •

"Another simple arithmetical one: Suppose you and I have the same number of copper coins. How many must I give you so that you have ten more than I?"

• 5 •

"A man once brought to a jeweler six chains of five links each. He wanted to make them all into one large, closed circular chain and asked the jeweler how much it would cost. The jeweler replied: 'Every link I cut open and then close costs a dollar. Since you want a circular chain and you have six chains, then it will cost you six dollars.'

" 'No,' said the man, 'the job can be done for less.'

"The man was right," said the Sorcerer. "Why?"

• 6 •

"My uncle once fooled me by telling me that a certain beggar had a brother and the brother died. But while the brother was alive, he never had a brother. How is this to be explained?"

• 7 •

"My uncle also told me about a very absentminded professor he knew who had three daughters. He once asked the professor how old his daughters were. The professor replied: 'I'm not too sure. I know that one of the three is the youngest.'

" 'That's not too surprising,' my uncle replied. 'Which one is the youngest?'

" 'I really don't know for sure; it's either Alice or Mabel.'

" 'Well, which one is the oldest?'

" 'I'm not sure about that either. I recall that either Alice is the oldest or Lillian is the youngest, but I can't remember which.' "

"Which one," asked the Sorcerer, "is the oldest and which the youngest?"

· 8 ·

"Another family problem my uncle told me: A certain boy had as many brothers as sisters. His sister Grace had twice as many brothers as sisters. How many brothers and sisters were there in the family?"

· 9 ·

"Here's a tricky one," said the Sorcerer. "If 5 cats can catch 5 mice in 5 minutes, how many cats are required to catch 100 mice in 100 minutes?"

· 10 ·

"Here is a simple problem my uncle told me that many people nevertheless get wrong. A certain miller took as toll one-tenth of the flour that he ground for a customer. How much did he grind for a customer who had just one bushel after the toll had been taken?"

· 11 ·

"My uncle once told me of an ancient puzzle propounded in
A.D. 310 by someone named Metrodorus. It concerns one
Demochares, who has lived one-fourth of his life as a boy,
one-fifth as a youth, one-third as a man in his prime, and
thirteen years in his advancing age. How old is he?"

· 12 ·

"My uncle once gave me a problem I spent many hours over.
Had I used my head, I would have realized the truth of the
matter in about a minute. The problem was this:

"We are given an 8 × 8 board divided into 64 squares, and
2 diagonally opposite corners are removed.

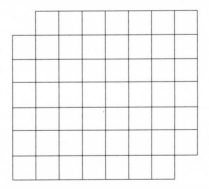

"We are given a pile of dominoes, and each domino can
cover 2 squares. The problem is to pave the surface. The *whole*
of each domino must be used—that is, each domino must lie on
2 squares. How can this be done?
"The principle involved in the solution provides a beautiful
example of what constitutes a mathematical *proof*."

• 13 •

"My uncle told a delightful incident about the American puzzle-master Sam Lloyd. Lloyd was also a magician, and he did one trick with his twelve-year-old son aboard ship that fooled even magicians. The boy was blindfolded and had his back turned to the audience. One of the spectators, not a confederate, took a deck of cards, shuffled it, and showed the faces one by one to the father. Each time, the boy correctly named the card. This was done before radio was invented, so radio signals are not the solution. As I said, even magicians were fooled. Of all the tricks I know, this one strikes me as the cleverest. How was it done?"

• 14 •

"Do you know of the logician Raymond Smullyan?" asked the Sorcerer.

"Never heard of him," replied Annabelle. "Who is he?"

"I've never heard of him either," said Alexander.

"No matter," replied the Sorcerer. "He invented many puzzles which my uncle was fond of. One of them is about a man who had two ten-gallon jars. Six gallons of wine was in one, and six gallons of water in the other. He poured three gallons of wine into the water jar and stirred; then poured three gallons of the mixture back into the wine jar and stirred; then three gallons of the mixture now in the wine jar into the water jar, and so on, back and forth, until the concentration of wine in the two jars was the same. How many pourings were necessary?"

• 15 •

"My uncle also told me of a nice variant Smullyan devised of the classical colored-hats problem. Do you know the colored-hats problem?"

"I think I once heard it, or something like it, but I don't really remember it," said Annabelle.

"I don't believe I've ever heard it," said Alexander.

"Well, there are many versions," said the Sorcerer. "One of the simpler ones is this: Three men—A, B, and C—are blindfolded and told that either a red or a green hat will be placed on the head of each. The blindfolds are then removed, so each of the three can see the colors of the hats of the other two, but neither can see the color of his own. Actually, all three are given green hats. The three are then asked to raise a hand if they see at least one green hat, so all three raise their hand. Then they are asked to lower their hand if they know what they have. After some pause the cleverest of the three lowers his hand. How does he know the color of his hat?"

• 16 •

"Now, here is Smullyan's variant. This time the three people—A, B, and C—are of equal intelligence; in fact all three are *perfect* logicians in that they can instantly deduce all consequences of any given set of premises. It is also common knowledge among the three that all three are perfect logicians. There are four red stamps and four green stamps available. While their eyes are closed, two stamps are pasted on the foreheads of each and the remaining two stamps are put into a drawer. After the three open their eyes and each sees the foreheads of the other two, A is asked whether he knows what he has, and he says no. Then B is asked if he knows what he has, and he says no. Then C is asked, and he says no. Then A is asked a second time, and he says no. Then B is asked a second time, and he says yes. How does B know what he has?"

• 17 •

"One of my uncle's favorites among Smullyan's problems was this: In 1918, on the day that the armistice of World War I was signed, three married couples celebrated by having dinner together. Each husband is the brother of one of the wives and each wife is the sister of one of the husbands; that is, there are three brother-sister pairs in the group. We are given the following five facts:

1. Helen is exactly twenty-six weeks older than her husband, who was born in August.
2. Mr. White's sister, who is married to Helen's brother's brother-in-law, married him on her birthday, which was in January.
3. Marguerite White is not as tall as William Black.
4. Arthur's sister is taller than Beatrice.
5. John is fifty years old.

"What is the first name of Mrs. Brown?"

Solutions

1. This problem is most easily solved by working it backwards. How many biscuits did the last dog find in order that by eating half of them and one more, all were gone? The only possibility is two. The dog before that must have found six; the dog before that must have found fourteen, and the first dog must have found thirty.

2. Some people guess that they are the same. This, of course, is wrong; six dozen dozen is 6 × 144, whereas half a dozen dozen is the same as six dozen (not six dozen dozen), which is

72. Actually, six dozen dozen is the same as half a dozen dozen dozen.

3. Let's say each item sells for a dollar wholesale and two dollars retail. Then, out of ten items sold, nine are sold for one dollar, yielding nine dollars, and one is sold for two dollars. So ten items bring in eleven dollars, two of which come from the retail stand. And so two-elevenths of his gross income comes from the retail stand.

4. The answer is 5, not 10!

5. What the jeweler had in mind was to open one end link of each of the six chains and then join them together to form a large circular chain. This would indeed cost six dollars. But each of the six chains had only five links apiece, and so one could cut open the five links of any one chain and use those five loose links to join the remaining five chains into a circle. Thus, the job could be done for five dollars.

6. The beggar was a woman.

7. Since either Alice is the oldest or Lillian is the youngest, it is impossible that Alice is the youngest, for if she were, it could neither be true that Alice is the oldest nor Lillian the youngest. Thus, Alice is not the youngest. Since either Alice or Mabel is the youngest, then it is Mabel who is the youngest. Therefore, Lillian cannot be the youngest, but either she is the youngest or Alice is the oldest, and since she is not the youngest, then Alice is the oldest. And so, Mabel is the youngest, Lillian is the middle one, and Alice is the oldest.

8. There were four brothers and three sisters.

9. The answer is five.

10. The common wrong answer is $1\frac{1}{10}$; the correct answer is $1\frac{1}{9}$. Let us check: $\frac{1}{10}$ of $1\frac{1}{9}$ is $\frac{1}{10} \times \frac{10}{9} = \frac{1}{9}$. So the man brought 1⅑ bushels to the miller, gave him 1/10 of it, which is ⅑ of a bushel, and kept the remaining 9/10 of 10/9 for himself, which is one bushel.

11.   Demochares must have been sixty. This can be found by trial and error, or by solving the algebraic equation $x/4 + x/5 + x/3 + 13 = x$, where x is his age.

12.   Imagine the squares colored alternately red and black like a checkerboard. The two corner squares that have been removed are of the same color—say, red. Thus, there are now more black squares than red squares—32 black squares and only 30 red squares. Now, a domino covers one black and one red square, so the total number of black squares covered must be the same as the total number of red squares covered. Since there are more black squares than red squares, a solution is impossible!

This is one of the simplest and neatest proofs of the impossibility of doing something that I know. Of course, if the corner squares were restored and any two squares of the same color were cut out, a solution would still be impossible. It has been proved (I hear) that if any two squares of *different* colors are removed, regardless of where they are, a solution is possible.

13.   The child never said a word; Sam Lloyd was a ventriloquist!

I have been a magician for many years and know all sorts of ingenious magical secrets. But this strikes me as the cleverest thing in magic that I know! The funny thing is that an elderly gentleman said to Sam Lloyd: "You shouldn't strain the boy's mind that much; it's not good for him!"

14.   It cannot be done in *any* finite number of pourings! To begin with, the concentration of wine in the wine jar is obviously greater than the concentration in the water jar. After the first pouring, the water jar is weaker (in wine) than the wine

jar. In the next pouring, the weaker is being poured into the stronger, and so the water jar is still weaker than the wine jar. Then some of the stronger is poured into the weaker, and so the water jar is still weaker than the wine jar. At any stage, the water jar is weaker in wine than the wine jar, and so at the next stage, regardless of which jar is poured into which, the water jar is still weaker than the wine jar.

Of course, this analysis is purely mathematical, assuming that wine is a totally homogeneous substance, rather than ultimately composed of discrete particles. As far as the real physical world is concerned, I don't know how many pourings would be necessary for the two concentrations to be *observably* indistinguishable.

15.   Let's say it was A who was the most intelligent. He reasoned this way: "Suppose my hat were red. Then B would know that if he too were red, C couldn't have raised his hand, indicating that he saw at least one green. And so if I were red, B would know that he is green. But B doesn't know that he is green. Hence, I can't be red; I must be green."

16.   The logic in this problem is far more intricate. To begin with, it is obvious that C did not see four stamps of the same color, since he would then have known what he had. Nor could C have seen two reds on one of A or B and two greens on the other, for suppose he saw two reds on B and two greens on A. Then, C would have known that he couldn't have two reds, since A would have then seen four reds and known what he had. Nor could C have two greens, since B would have then seen four greens and known that he had two reds.

This proves that it is not possible that both of A's stamps are the same color *and* that both of B's stamps are also the same color; at least one of the two has to have red and green stamps. A and B both realized this after C said that he didn't know what he had, and each realized that the other realized this.

Therefore, on the second round, B realized that if his colors were the same, then A would have known on the second round that he was red/green, which he didn't. And so, B realized that he must himself be red/green.

17. *Step 1.* Since Mr. White's sister is married to Helen's brother's brother-in-law, then she is not married to Helen's brother. Also, Marguerite is Mrs. White. Using these two facts, we will see that Mr. White's sister must be Helen.

Now, Helen's brother is married either to Marguerite or to Beatrice. If it is Marguerite, then Helen's brother is Mr. White (since Mr. White is married to Marguerite), in which case Helen is Mr. White's sister. On the other hand, suppose it is Beatrice to whom Helen's brother is married. Then Beatrice is not Mr. White's sister (because Mr. White's sister is *not* married to Helen's brother), and since Marguerite is not Mr. White's sister (she is his wife) then it must again be Helen who is Mr. White's sister. This proves that Helen is Mr. White's sister.

Step 2. Helen is either Mrs. Brown or Mrs. Black (since Marguerite is Mrs. White). We now show that if Helen is Mrs. Brown, then she must be married to John, which we will later show cannot be.

Well, suppose Helen is Mrs. Brown. Then Beatrice is Mrs. Black (since Marguerite is Mrs. White). Hence Beatrice is not Mr. Black's sister, nor is she Mr. White's sister (since Helen is). Hence Beatrice is Mr. Brown's sister. Now, Arthur's sister is not Beatrice (she is taller than Beatrice), but Mr. Brown's sister is Beatrice; hence Arthur is not Mr. Brown. He is also not Mr. Black (William is); hence Arthur is Mr. White. Then John, who is not Mr. White nor Mr. Black (William is), must be Mr. Brown. Then, by our assumption that Helen is Mrs. Brown, Helen must be married to John.

This proves that if Helen is Mrs. Brown, she is married to John.

Step 3. But Helen *can't* be married to John for the following interesting reason:

We have already proved that Helen is Mr. White's sister (Step 1). And we have been given that Helen (as Mr. White's sister) was born in January. Thus Helen was born in January, her husband was born in August, and Helen is exactly twenty-six weeks older than her husband. Using a calendar, we can see that the only way this is possible is that Helen was born on January 31 and her husband was born on August 1, *and there is no February 29 in between!* Therefore, Helen and her husband were not born in a leap year. But John, being fifty in 1918, *was* born in a leap year. Therefore John cannot be Helen's husband.

But we saw (Step 2) that if Helen were Mrs. Brown, then John would be Helen's husband. Since he is not, then Helen can't be Mrs. Brown. Also, Marguerite is not Mrs. Brown (she is Mrs. White), and so it is Beatrice who is Mrs. Brown.

9

THE PLANET OG

"Did your uncle invent any puzzles of his own?" asked Alexander.

"Heavens, yes! A very great many," replied the Sorcerer.

"Did he publish any of them?" asked Alexander.

"You remind me," said the Sorcerer, "of the story of the two American professors standing in Florence before a statue of Jesus. 'Now, there was a great teacher!' said one. 'Yes,' said the other, 'but he never published!'

"This obsession with publishing! No, my uncle never published any of his problems; unfortunately, he never even wrote them down. He would improvise them on the spot—mainly, I think, to amuse me. Then he would forget them completely. Fortunately, I have remembered many of them."

"Will you tell us some?" asked Annabelle.

"Surely," replied the Sorcerer. "One particular group occurs to me now. They are all about a planet my uncle called *Og.*"

"Is there really such a planet?" asked Alexander.

"I doubt it," replied the Sorcerer. "I'm pretty sure he made the name up. Anyway, I think he was influenced by science fiction, particularly by *The Princess of Mars*, a book by Edgar Rice Burroughs. As in Burroughs's Martian fantasy, my uncle's planet Og is inhabited by two races—the green people and the red people. Also, the people from the northern hemisphere— that is, those who were born there—are very different from the southerners, those born in the southern hemisphere."

"How are they different?" asked Alexander.

"Well," replied the Sorcerer, "the curious thing about this planet is that the green northerners always tell the truth and the red northerners always lie, whereas the green southerners lie and the red southerners tell the truth."

"Oh, dear!" said Annabelle, "I can see that these puzzles are going to be very complicated!"

"Some are complicated and some are simple," replied the Sorcerer. "Let's start with a simple one."

Here are some of the puzzles the Sorcerer recalled.

1 • One Dark Night

Suppose you meet an inhabitant on the street one dark night and cannot see whether he is red or green. Also, you don't know from which hemisphere he hails. What single yes-no question could you ask him that would determine his color?

2 • Another Dark Night

Once a traveler from our planet Earth visited the planet Og and met a native on the street one very dark night.

"Are you red?" he asked. The native didn't answer.

"Are you a southerner?" the traveler asked. Again the native didn't answer.

"Are you not going to say *anything?*" asked the traveler.

Before walking away the native replied: "If I answered no to both of your first two questions, I would be lying at least once."

Is it possible from this to determine the native's color and what hemisphere he is from?

3 • Another Dark Night

On another dark night, a traveler from Earth met a native and asked him: "Are you red?" The native said that he was. Then the stranger asked him from which hemisphere he came. The native replied, "I refuse to tell you," and walked away.

From which hemisphere was he?

4 • In Broad Daylight

One bright day, a traveler from Earth met an Og native who said, "I am a green northerner." The traveler, though good at logic, could not deduce whether the native was a northerner or a southerner (though, of course, he saw what color he was).

What color was the native?

5 • A Point of Logic

There is a difference between a native of Og saying "I am a green northerner," and his making two separate statements— one that he is green, and the other that he is a northerner. In the one case, it cannot be deduced what the native really is, but in the other case, it can be. In which case can it be, and in that case, what is the native?

6 • Another Question of Logic

Suppose a native says: "If I am green, then I'm a southerner."
Can it be deduced what color he is? Can it be deduced from which hemisphere he hails?

7 • A Duo

This problem concerns two inhabitants of Og, a red northerner and a southerner (possibly red and possibly green). One of the two is named Ark and the other Bark. They made the following statements:

Ark: Bark and I are the same color.

Bark: Ark and I are different colors.

Which one is right? Which one is the northerner? What color is the southerner?

8 • Another Duo

Two natives named Ork and Bork made the following statements:

Ork: Bork is from the north.

Bork: Ork is from the south.

Ork: Bork is red.

Bork: Ork is green.

What color is Ork and where is he from? And Bork?

9 • Two Brothers

Any two siblings on the planet Og are necessarily the same color, but not necessarily from the same hemisphere. (The mother might have given birth to one and then, sometime later, crossed the equator and given birth to the other.) But they definitely *are* the same color.

Two brothers, Arg and Barg, once made the following statements about themselves:

Arg: We were born in different hemispheres.

Barg: That is true.

Were they lying or telling the truth?

10 • Brothers?

In this more complicated situation, two natives, Org and Borg, made the following statements:

Org: Borg is from the north.
Borg: In fact both of us are from the north.
Org: That is not true!
Borg: Org and I are brothers.

Are the two really brothers? What color is each? Where is Org from? What about Borg?

11 • A Trial

A native was being tried for robbery. It was important for the court to decide whether he was a northerner or a southerner, since it was known that the theft was committed by a northerner. The defense attorney did not want his own color to be known, so he appeared masked and gloved in court. To everyone's surprise, he claimed that he and the defendant were both northerners. (One would have expected him to have made a statement proving the defendant's innocence!) To everyone's even greater surprise, when the prosecutor cross-examined the defense attorney, the latter claimed that he himself was not a northerner.

How do you explain this? Can the color of the defense attorney be established? Can it be determined whether the defendant was guilty or innocent?

12 • What Are They?

Two inhabitants, A and B, were of different colors and from different hemispheres. They made the following statements:

A: B is a northerner.

B: A is red.

What colors are A and B and where are they from?

13 • What Color Is Snarl?

A southerner named Snarl once claimed that he overheard a conversation between two brothers, A and B, in which A said that B was a northerner and B said that A was a southerner.

What color is Snarl?

14 • Another Duo

Two natives, A and B, of different colors made the following statements:

A: B is a northerner.

B: Both of us are northerners.

What are A and B?

15 • Is There a King?

The anthropologist Abercrombie once visited the planet Og and was curious to find out if the planet had a king. He had the following conversation with a native:

Abercrombie: I was told that you once claimed that this planet has no king? Is that true?

Native: No, I never claimed that!

Abercrombie: Well, have you ever claimed that this planet *does* have a king?

Native: Yes, I have claimed that.

Can it be determined whether the native was truthful or lying? Can it be determined whether or not the planet has a king?

16 • Is There a Queen?

Abercrombie was also interested in finding out whether the planet has a queen. He found the answer when he came across two brothers, A and B, who made the following statements:

A: I am a northerner, and this planet has no queen.

B: I am a southerner, and this planet has no queen.

Is there a queen or not?

17 • What Color Is the King?

This puzzle is one of the Sorcerer's favorites.

The King of Og always wears a mask and gloves, and none of his subjects ever learns his color. His brother, the Duke of Snork, is similarly disguised.

One day, Abercrombie came to the court and the King and his brother decided to test their visitor's intelligence. First, Abercrombie had to swear that if he should find out the color of the King, he would never tell any of the inhabitants. Then he was ushered into a room in which the King and his brother were both seated—both masked and gloved, of course—and one of them said: "If I am a green northerner, then I am the King." The other one then said: "If I am either a green northerner or a red southerner, then I am the King."

What color is the King?

Solutions

1. A question that works is: "Are you a northerner?" If he answers yes, he is green; if he answers no, he is red. I leave the proof to the reader.

Of course, the traveler could equally well have found out if the native was a northerner or a southerner by asking: "Are you green?" The situation has the pretty symmetry that to find out if the native is green, you ask him if he is northern, and to find out if he is northern, you ask him if he is green.

2.   The native said in effect that he is *either* red *or* southern (and possibly both)—in other words, that he is not a green northerner. A green northerner would not lie and say that. Also, neither a green southerner nor a red northerner could make the truthful statement that he is not a green northerner. And so the native must have been a red southerner.

3.   When the native said "I refuse to tell you," he spoke truly, since he really did refuse to tell the traveler. And so the native is truthful. Hence, his first answer was truthful; so he really is red. Since he is truthful and red, he must be from the south.

4.   If the native had been red, the traveler would have known that he was a northerner (since a red southerner would never claim to be a green northerner), but since he didn't know, then the native must have been green (possibly a truthful green northerner or possibly a lying green southerner).

5.   In the case of a native who claims to be a green northerner, nothing can be deduced other than that he is not a red southerner. But it is a very different story if a native first claims to be green and then makes a separate claim that he is a northerner. For suppose he makes these two separate claims. Then, they are either both true or both false. If they are both false, then he must be red (since the first claim is false) and he must be a southerner (since the second claim is false). It is impossible for a red southerner to make false claims; therefore, both claims must be true. So the native is a green northerner.

6.   To say that if he is green then he is a southerner, is tantamount to saying that he is not a green northerner. And so the native, in effect, claimed that he was not a green northerner.

Only a red southerner could make such a claim. (This is really the same as Problem 2; only the wording is different.)

7. Since the two disagree, one is telling the truth and the other is lying. The red northerner must be lying; hence the southerner is telling the truth, and therefore he must be a red southerner. So the two really are the same color; which means Ark told the truth and Bark lied. So Ark is the red southerner and Bark is the red northerner.

8. If Ork's statements are true, then Bork is a red northerner, and if Ork's statements are false, Bork is a green southerner. In either case Bork is a liar. Thus, Bork's statements are both false, and Ork is a red northerner. Hence Ork lied, so his statements are both false and Bork is a green southerner.

9. If it were really true that they were from different hemispheres, then we would have the impossible situation of a northerner and a southerner *of the same color* agreeing with each other, which is impossible. So what they said was not true.

10. Since Org contradicted Borg, one is lying and the other is telling the truth. Suppose Borg is telling the truth. Then both are northerners (by Borg's first statement) and both are brothers (by Borg's second statement); hence both are the same color and we have the impossible situation of two northerners of the same color disagreeing. Therefore, Borg was not telling the truth; he is a liar, and it is Org who is truthful.

Then Borg is a northerner as Org truthfully said, but they are not both northerners (since Borg lied and claimed they were); hence Org is a southerner. Thus Org is a truthful southerner and must be red and Borg is a lying northerner and also red. Org and Borg are *not* brothers (since Borg said they were), even though they are the same color.

11. Obviously, the defense attorney's claims cannot both be true, hence he is a liar. Since he denied being a northerner, he really is a northerner, and must be a red northerner. Since his claim that he and his client are both northerners was false and he is a northerner, then the defendant must be a southerner. And so the defendant is not guilty.

The defense attorney, despite the strangeness of his behavior, was not so stupid; he got his client acquitted.

12. Since A and B are different in color and are also from different hemispheres, they must either both be telling the truth or both lying. Suppose their statements were both false. Then, B would be southern and A would be green. So B would be red (since A is green) and also A would be northern (since B is southern). Hence we would have a red southern B (as well as a green northern A) making false statements, which is not possible. Therefore, both statements must be true. Hence B is a northerner and A is red, and so B is a green northerner and A is a red southerner.

13. If Snarl's account were true, we would have the following contradiction: Since A and B are brothers, they are the same color. Suppose they are red. If A is a northerner, he is a red northerner; hence his statement is false and B is really a southerner, a red southerner who is therefore truthful and couldn't have made the false statement that A is a southerner. On the other hand, if A is a southerner, he is a red southerner; hence his statement is true and B really is a northerner, a red northerner, therefore untruthful. Yet he would truthfully have said that A is a southerner. This is impossible; hence the brothers cannot be red. A symmetrical argument, which we leave to the reader, reveals that they cannot be green either.

Another way of looking at it (which I believe is more instructive) is to first realize that in any land in which every inhabitant is either a constant liar or a constant truth-teller, it is impossible to have inhabitant X saying that inhabitant Y is lying and Y saying that X is truthful (for then Y is assenting to X's claim that Y lies, which neither a truth-teller nor a liar can do). Now, in the present problem, since A and B are the same color, if A calls B a northerner and B calls A a southerner, then one of the two is effectively calling the other a liar and the other is effectively calling the first truthful. (If they are red, A is effectively calling B a liar and B is effectively calling A truthful. If they are green, we have the reverse.) But this cannot be.

At any rate, Snarl's account was false, and since he is a southerner, he must be green.

14. Suppose B told the truth. Then the two are both northerners; hence A's statement that B is a northerner is true. We then have the impossibility of two northerners of *different colors* both telling the truth. Thus B did not tell the truth, so at least one of them is a southerner. Suppose B is a northerner. Then A must be the southerner. Also, A told the truth that B is a northerner, hence A must be a red southerner and B would be green, and we would have the impossibility of a green southerner telling the truth. Therefore, B is not a northerner but a southerner. Since B is a southerner and lied, he must be a green southerner. Also, since B is a southerner, A lied, and A is red (since B is green). Thus, A is a red northerner.

So A is a red northerner and B is a green southerner, and both lied.

15. It cannot be determined whether the native was truthful or lying, but it can be determined whether or not there is a king.

Suppose the native is truthful. Then he really did once claim that the planet had a king and his claim must have been true, so the planet does have a king.

On the other hand, if the native is a liar, then it is not true that he never claimed there was no king; hence he actually did once claim there was no king, and being a liar, this claim was false. Hence there must be a king.

Thus, regardless of whether or not the native is truthful, the planet does have a king.

16. The two brothers are the same color. Suppose they are red. Then A's statement can't be true, since if it were, then A would have to be a northerner (as he claimed) and we would have a red northerner telling the truth, which cannot be. Therefore, A's statement is false. Thus A is red and makes false statements; so A is a northerner. Therefore, if there were no queen on the planet, then it would be true that A is a northerner and there is no queen; hence A's statement would be true after all, which it isn't! This proves that if the brothers are red, then the planet must have a queen.

By a symmetrical argument, using B's statement instead of A's, if both brothers are green, then the planet has a queen. And so the planet has a queen.

17. Let A be the first one who spoke and B the second. Is A's statement necessarily true? That is, if A is a green northerner, must he then necessarily be the King? Well, suppose he *is* a green northerner. Then, he is truthful, so it really is true, as he said, that if he is a green northerner, he is the King. This proves that if he is a green northerner, then he is the King. Since he said just that, he is truthful. So now we know that A is truthful. From this it doesn't follow that A must be the King; all we know is that if he is a green northerner, then he is the King. But we don't know whether or not he is a green northerner. (We know that he is truthful, but he could be a red southerner.)

Now, with B's statement it is different. Again, B's statement must be true, because if he were either a green northerner or a red southerner, his statement would have to be true. Hence it would be true that if he is either a green northerner or a red southerner, he would have to be the King. And so, if he is either a green northerner or a red southerner, then he is the King. Since he said just this, he is truthful; hence he must be either a green northerner or a red southerner. But we have already seen that if he is either a green northerner or a red southerner, then he must also be the King. And this proves that he is the King.

Since B is the King, then A is not; hence A cannot be a green northerner (because if he were, then he *would* be the King, as we have shown). Since A is not a green northerner but is truthful, then A must be a red southerner. Since A is red, so is his brother, the King. Thus the King is red (and like his brother, a red southerner).

10
METAPUZZLES

• 1 •

"Here is another puzzle about the planet Og. It is of a very special type," said the Sorcerer. "It might aptly be called a *metapuzzle.*

"A native on this planet claims to be a married northerner. If I told you what color he is, would you have enough information to deduce whether or not he is married?"

Annabelle and Alexander went to work on this, and sometime later Alexander said: "We don't know; there is no way of telling."

"You are right," said the Sorcerer, "and now I will tell you that if I told you the native's color, you would then have enough information to be able to determine the native's marital status."

"Great!" said Annabelle. "Now I know whether or not the native is married."

Is the native married or not?

2 • Another Metapuzzle

"Here is another metapuzzle about the planet Og," said the Sorcerer.

"A logician from our planet visited Og and met a native one dark night and asked him whether he was a green northerner. The native answered (yes or no), but the logician couldn't tell from his answer what he was.

"A second logician met the native on another dark night and asked him whether he was a green southerner. The native answered (yes or no), but the logician couldn't figure out what he was.

"On another dark night, a third logician met the native and asked him whether he was a red southerner. The native answered (yes or no), but the logician couldn't figure out what he was.

"What was he?" asked the Sorcerer.

3 • A Metametapuzzle

"I loved the last two puzzles," said Annabelle, "but why do you call them *metapuzzles?* Just what does the word mean?"

"The term was invented by my uncle," replied the Sorcerer. "Metapuzzles are, so to speak, puzzles about puzzles. One solves a metapuzzle on the basis of knowing that a certain other puzzle, or puzzles, can or cannot be solved. Such puzzles can be quite intricate!

"Some puzzles go one level deeper and can be solved only by knowing whether certain metapuzzles can be solved! Such puzzles my uncle called *metametapuzzles*. My uncle was a master at inventing these."

"Can you give us an example of a metametapuzzle?" asked Alexander.

The Sorcerer thought a moment. "All right," he said, "my

uncle once gave me a metapuzzle about a logician who visited the planet Og and met a native one dark night and was curious to know whether or not the native was truthful. He then asked the native which of the four types he was (green northerner, red northerner, green southerner, red southerner) and the native then named one of the four types and claimed to be that."

"Oh," said Annabelle, "was the logician then able to determine whether or not the native was truthful?"

"Good question," replied the Sorcerer, "and I asked my uncle just that."

"Did he answer you?" asked Alexander.

"Yes."

"What answer did he give?" asked Alexander.

"I'm not telling you."

"Well then," said Annabelle, "after your uncle told you whether or not the logician could solve *his* problem, were you then able to determine whether or not the native was truthful?"

"I'm not telling you that either," replied the Sorcerer. "If I told you that, then *you* would be able to determine whether or not the native was truthful."

"Then why won't you tell us?" asked Alexander. "Don't you want us to solve the problem?"

"It's no longer necessary," said the Sorcerer with a smile. "You have enough information to determine whether or not the native was truthful."

At this point, the reader has been given enough information to answer the following questions:

1. Was the native truthful?
2. Was the Sorcerer able to solve the metapuzzle given him by his uncle?
3. Could the Sorcerer determine which of the four types the native was?

4. Did the logician determine whether or not the native was truthful?

5. Could the logician determine which of the four types the native was?

I believe the reader will find this puzzle quite instructive!

4 • A Knight-Knave Metapuzzle

"That was quite ingenious," said Annabelle. "Have you ever invented any metapuzzles or metametapuzzles yourself?"

"Oh yes," said the Sorcerer. "I learned the art from my uncle. I have even invented metametametapuzzles—and gone even above these. I have constructed puzzles of all degrees of complexity ranging from very simple to very difficult."

"Are they also about the planet Og?" asked Annabelle.

"Oh no. They are mainly about ordinary knight-knave islands like this one—islands in which all knights tell the truth and all knaves lie and every inhabitant is either a knight or a knave.

"Let's see now," the Sorcerer continued, "here's one you might like: A logician visited a knight-knave island and met two inhabitants, A and B, and asked A: 'Is it true that B once said that you are a knave?' A answered (yes or no). Then the logician asked one of the two whether the other was a knave. He was answered (yes or no). It is not given whether or not the logician could then tell what they were.

"A second logician came across the same pair on another occasion and asked A whether B had once claimed that A and B were both knaves. A answered (yes or no). Then the logician asked one of them whether the other was a knave, and was answered (yes or no). It is not given whether or not the second logician could then solve the problem.

"What *is* given is that one of the two logicians was able to solve the problem and the other wasn't, but we are not told which one solved it. And now *your* problem is to determine what A and B are, and which logician solved the problem."

5 • A Court Case

"Wow! That was a hard one," said Annabelle, after the two had solved it, "but quite intriguing. Do you have any more?"

"The next is a metametapuzzle," said the Sorcerer. "It concerns a court trial. A difficult feature of this trial was that it was not known whether the prosecutor was a knight or a knave, nor what the defense attorney was. Anyway, the two made the following statements:

Prosecutor: The defendant is guilty and has committed other crimes in the past.

Defense Attorney: The defendant is innocent, and the prosecutor is a knave.

"The judge then asked a question whose answer he already knew, and one of the two attorneys answered it, thus revealing to the judge whether he was a knight or a knave, and the judge then knew whether the defendant was innocent or guilty.

"A highly intelligent reporter was present at the trial and later told a logician friend of his all I have told you so far and no more. The logician worked on the problem for a while and realized he did not have enough information to solve it. 'Tell me,' he said to the reporter, 'if the other attorney had answered the question instead, would the judge then have known whether the defendant was guilty?' The reporter thought for a while and said: 'I really don't know. In fact, there is no way of knowing.'

" 'Oh, good!' said the logician. 'Now I know whether the defendant was innocent or guilty.'

"Was the defendant guilty or innocent? Also, which attorney answered the judge's question, and was he a knight or a knave?"

6 • A Very Abstract Hyper-Metapuzzle

"I recently thought of a *very* abstract metapuzzle," laughed the Sorcerer. "Actually, the solution is not at all complicated; in fact, it is very simple. Still, it is tricky and one is apt to overlook it.

"By a *perfect logician*, I mean one who, given any problem, if he is given enough information to solve the problem, will then know that there is enough information and will solve the problem, but if he is not given enough information, then he will know that it is not enough. Now, a certain perfect logician was given a certain problem—call it Problem P. I am not telling you what Problem P is or whether he was given enough information to solve it. All I will tell you is that if I *did* tell you what Problem P was and whether or not the logician had enough information to solve it, then you would have enough information to solve Problem P. By *you*, I of course mean anybody. That is, the information of what Problem P is, together with whether the logician had enough information to solve it—that information is enough to solve Problem P.

"And now *your* problem is: Did the logician solve Problem P or didn't he?"

"The time has come," said the Sorcerer, "for a complete change of pace. I suggest that we visit the Island of Robots. I believe that it will interest you enormously! Tomorrow is supposed to be an unusually clear and beautiful day, and I think we should leave before sunrise so we can have an entire day on the island. Why don't you stay the night so that we can leave really early?"

This sounded like an attractive idea, and the couple happily agreed. What happened on the Island of Robots will be told in the next chapter.

Solutions

1. The point is that if the native is red, he cannot be married, but if he is green, he might or might not be married. Here is why.

Suppose he is red. If he is truthful, he is a married northerner (as he claimed); hence he is a red northerner who is truthful, which cannot be. Therefore (assuming he is red) he is not truthful; hence he is a red northerner, but not a married northerner (since he falsely claimed he was). This proves that if he is red, he must be unmarried.

On the other hand, if he is green, then he could be a green married northerner who speaks truly, or a married or unmarried green southerner who speaks falsely. Thus, the knowledge that he is green does not solve the question of whether he is married, but the knowledge that he is red does. Now, the Sorcerer said that the knowledge of the native's color *would* solve the question of his marital status, and so it must be that the native is red and unmarried.

2. Let us label the questions asked of the native.

Q1. Are you a green northerner?

Q2. Are you a green southerner?

Q3. Are you a red southerner?

As the reader can easily check, here are the answers that each of the four types would give if asked the question.

	Q1	Q2	Q3
Green northerner	Yes	No	No
Green southerner	Yes	No	Yes
Red northerner	Yes	Yes	Yes
Red southerner	No	No	Yes

In each case, three would have answered one way and one the other. Had the first answer been no, the logician would have known that the native was a red southerner, but since he didn't know, the native didn't answer no and hence was not a red southerner. Also the native wasn't a red northerner (since the second logician didn't know) and he wasn't a green northerner (since the third logician didn't know). Therefore, the native was a green southerner.

3. We leave it to the reader to verify the following four simple facts.

(1) If a native claims to be a green northerner, then he can be a green northerner, a red northerner, or a green southerner (but he can't be a red southerner).

(2) If a native claims to be a red northerner, he can be only a green southerner.

(3) If he claims to be a red southerner, then he can be a red southerner, a green southerner, or a red northerner (but he can't be a green northerner).

(4) If he claims to be a green southerner, then he can be only a red northerner.

Therefore, if the native had claimed either to be a green northerner or a red southerner, the logician would have had no way of knowing which of the three types he might be, hence the logician couldn't have known whether the native was truthful.

On the other hand, if the native claimed either to be a red northerner or a green southerner, the logician would have known that the native was lying—indeed, he would have even known whether the native was a red northerner or a green southerner (whichever he claimed to be, he must really be the other). And so in short, if the logician was able to determine the native's veracity, the native must have lied, but if the logician couldn't determine his veracity, then the native might have lied or he might have told the truth—there is no way of knowing.

Now, the Sorcerer's uncle told him whether or not the logician succeeded in his quest (to determine whether or not the native was truthful). If the uncle answered affirmatively, then the Sorcerer would know that the native must have lied, but if the uncle answered negatively (if he had told him that the logician didn't know), then the Sorcerer would have no way of knowing whether or not the native was truthful (all he would have known is that the native either claimed to be a green northerner and was thus anything but a red southerner) or claimed to be a red southerner (and was thus anything but a green northerner).

At this point, we know that if the Sorcerer solved the metapuzzle that his uncle gave him, then the native must be a liar, but if the Sorcerer didn't solve his metapuzzle, then there is no way of telling whether the native was a liar. But the Sorcerer further told us (or rather his two students) that a knowledge of whether the Sorcerer solved his metapuzzle would be enough to determine the veracity of the native. Well, the knowledge that the Sorcerer *didn't* solve his metapuzzle would not be enough, whereas a knowledge that he *did* solve it would be. And so the Sorcerer must have solved his metapuzzle and the native must have lied.

In summary, the native lied. The Sorcerer knew that the native lied, but couldn't determine which of two types of liar he was, whereas the logician not only knew that the native lied, but also whether he was a red northerner or a green southerner (though we have no way of knowing which).

4. *Step 1.* Let us see what we can learn from the first logician's encounter. A answered either yes or no to the logician's first question.

Case A1. He answered yes. Thus A affirmed that B had once called him a knave. If A is a knight, then B did once claim that A was a knave and B must be a knave. Thus A and B could not both be knights. On the other hand, if A is a knave, then B could be either, and there is no way of telling which (since B then made no such claim, nothing can be deduced about B). And so all the logician would know at this stage was that A and B were not both knights.

Case A2. A answered no. Thus A denied that B once called him a knave. Of course, A could be a knight and B could be either a knight or knave, but if A is a knave, then B must be a knight (because if A's denial is false, then B really did once call A a knave; hence B must be a knight). And so, in this case, all the logician would know at this stage is that A and B were not both knaves.

Step 2. Now, the answer to the logician's second question would reveal whether A and B were the same type, because if one native affirms that the other one is a knave, they must be different types (a knight would not claim another knight to be a knave, nor would a knave claim another knave to be a knave), and if one native affirms another to be a knight, the two must be of the same type. And so, after the second question was answered, the logician would then know whether the two were alike (of the same type). If he found out that they were alike, then he would have solved the problem of what each was, regardless of whether Case A1 or Case A2 held, because in Case A1, he would have known that they were alike but not both knights, hence both knaves, and in Case A2, he would have known they were both knights. So, regardless of what A's first answer was, if the second answer revealed the two to be alike, then the logician could solve the problem. On the other hand, if the second answer revealed that A and B were different types, then in neither Case A1 nor Case A2 could the logician have known which of A or B was the knight and which the knave.

We have thus proved:

(1) If the two are the same type, then the first logician has solved the problem.
(2) If the two are different types, then the first logician has not solved the problem.

Step 3. We now consider the second logician's encounter. This time, A, in answering the first question, either affirmed or denied that B claimed that they were both knaves.

Case B1. Suppose A affirmed that B claimed that they were both knaves. If A was a knight, then B must have been a knave. If A was a knave, B could be either. So all that the second logician would know in this case is that A and B were not both knights.

Case B2. Suppose A denied that B claimed that they were both knaves. Then A couldn't be a knave, for if he were, then B really did once claim that both were knaves, which is not possible (because if B were a knight, he wouldn't falsely make such a claim, and if B were a knave, he would never claim the true fact that both are knaves). And so, A must be a knight; B never did make such a claim and B could be either. Thus, if Case B2 is the one that occurred, all the second logician would know (and he would know this) is that A is a knight.

Step 4. Now, suppose after the logician's second question was answered, he knew whether or not A and B were the same type. Suppose he found that they were the same. Then, he must have solved the problem (in Case B1 he would know that both are knaves, and in Case B2 he would know that both are knights). Suppose, on the other hand, he found out that they were different. Then in Case B1, he couldn't solve the problem (he would know that not both were not knights and also that they were different, but he couldn't know which was which), whereas in Case B2, he would know that A was a knight and B a knave (since he would know that A is a knight and the two are different). This proves that if A and B are different, then the only way he could solve the problem is for A to be a knight and B a knave (because if A and B are different, Case B1 couldn't hold or the logician wouldn't have solved the problem, as we have seen), hence Case B2 must hold and therefore the logician knew that A is a knight and B a knave.

We have thus proved the following:

(3) If A and B are the same type, the second logician has solved the problem.

(4) If A and B are different types, then the only way the second logician could solve the problem is that A is a knight and B is a knave.

Step 5. Now we have all the necessary pieces! Suppose A and B are the same type. Then by (3), the second logician solved the problem and by (1) the first logician solved the problem, and this is contrary to the given condition that one and only one of the logicians solved the problem. Therefore A and B are different types.

Since A and B are different types, then the first logician didn't solve the problem, by (2); hence the second logician did. Hence by (4), A is a knight and B is a knave.

5. We must consider all four possible cases—which of the two attorneys answered the question and whether each revealed himself to the judge as a knight or a knave.

Case 1. The prosecutor revealed himself to be a knight. Then, of course, the judge knew that the defendant was guilty (and also that the defense attorney was a knave).

Case 2. The prosecutor revealed that he was a knave. Then the judge couldn't know what the defendant was (he could be innocent, or he could be guilty but, contrary to the prosecutor's claim, he might never have committed any crimes in the past).

Case 3. The defense attorney revealed that he was a knight. Then, obviously, the judge would know that the defendant was innocent (and also that the prosecutor was a knave).

Case 4. The defense attorney revealed himself to be a knave. Then the judge would know that the defendant must be guilty as follows: Since the defense attorney's statement is false, then it cannot be that the defendant is innocent *and* that the prosecutor is a knave; hence either it is false that the defendant is innocent or it is false that the prosecutor is a knave. If the former, then of course the defendant is guilty. If the latter, then the prosecutor is a knight, in which case the defendant is again guilty (since the prosecutor said he was). This proves that if the defense attorney revealed himself to be a knave, the judge would know that the defendant must be guilty (though he wouldn't know whether the prosecutor was a knight or a knave).

Now let's see why the reporter didn't know the answer to his friend's question.

The reporter was present at the trial, so he knew which attorney answered the judge's question and whether he revealed himself as a knight or a knave. Suppose it was the prosecutor who answered. Then, had the defense attorney answered instead, the judge *would* have solved the case (as we saw in our analysis of Cases 3 and 4), and so the reporter would have answered yes to his friend's question. Since he didn't answer yes, but instead said that he didn't know, then it was not the prosecutor who answered the judge's question. It was the defense attorney who answered and revealed his type.

Suppose the defense attorney was a knight. Then the prosecutor must be a knave, and therefore had he, instead of the defense attorney, answered the judge, the judge *wouldn't* have been able to solve the case (as we saw in our analysis of Case 2) and the reporter would know this and would have answered no to his friend's question.

This leaves only the possibility that the defense attorney revealed himself to be a knave. Then the reporter could not have told whether the prosecutor was a knight or a knave, and hence did not know the answer to his friend's question. (All he knew was that if the prosecutor had been the one to have answered the judge, and if he had revealed himself to be a knight, the judge could have solved the case, and if he had revealed himself to be a knave, the judge couldn't have. But the reporter had no way of knowing whether the prosecutor would have revealed himself to be a knight or a knave.)

6. Obviously, the logician knew what Problem P was, and since he was perfect, he knew whether or not he had enough information to solve the problem. This information is enough to solve the problem; hence the logician had enough information to solve the problem and, being perfect, he solved the problem.

PART III
SELF-REPRODUCING ROBOTS

11

THE ISLAND OF ROBOTS

THE SORCERER and his friends Annabelle and Alexander retired early and rose in the wee hours of the morning and set sail for the Island of Robots. They arrived just about as the sun was rising and spent the time before lunch walking around observing the strange things that went on. This island was the noisiest place that Annabelle or Alexander had ever been to in their lives. The clinging and clanging and binging and banging and din and clatter were almost unbearable. And the things that went on! The whole island was bustling with metallic robots moving about, some apparently aimlessly, others creating new robots from various parts lying around, and others dismantling robots whose parts would often be used for creating other robots. Each robot bore a label consisting of a string of capital letters. Annabelle and Alexander at first thought the letters were merely for identification, but later found they were a *program* determining what the robot should do—whether it would walk around aimlessly or whether it would create other robots, and if so, what program the new robots would have, or whether it would be a destructive robot, and if so, what robots it would destroy.

One thing struck the couple as quite peculiar: They saw a robot pick up a bunch of pieces and construct a robot that

107

looked identical to its creator; indeed it had the same program. Having the same program, this duplicate constructed a duplicate of *itself*, which in turn constructed a duplicate of itself, which in turn constructed a duplicate of itself (as a matter of fact, from parts of its great-grandparent, which had been dismantled by then). Indeed, this process could go on forever, unless something stopped it.

Next, the couple saw another curious thing: a robot—call it x—constructed a robot y, quite different from x, which then constructed a duplicate of x, which then constructed a duplicate of y, and this 2-cycle could go on forever!

Then they saw a 3-cycle: x constructed y, which constructed z, which constructed (a duplicate of) x, which constructed (a duplicate of) y, . . .

Next they saw something very upsetting. A robot x constructed a robot y and the first thing y did when it was completed was to destroy its creator. This struck the couple as the height of ingratitude.

Then they saw a robot x destroy a robot y. Later on, y got reassembled and met x and immediately destroyed it. Then x got reassembled sometime later and destroyed y, and this could go on forever!

Next they saw two very different-looking robots, x and y, walk briskly toward each other and immediately begin dismantling each other, and soon all that was left was a rubble of parts.

Then they saw a robot x construct a robot y, which constructed a robot z, which then dismantled x. Thus x was destroyed by its own grandchild!

Next, they saw something quite sad: a suicidal robot dismantled itself, and all that was left was a pile of parts. (At a certain stage of dismantling, the robot pressed a button which caused what was left to fall completely apart.) Then another robot came by and reassembled the dismantled robot, but when the second robot left, the first one, still having the same program, destroyed itself again. There seemed to be no hope for this poor robot. If it were ever again reassembled, still with the same program, then it would have to destroy itself again; no robot with this program could possibly be stable.

"I'm completely bewildered by all this," said Annabelle. "I have no idea what in the world is going on!"

"Nor I," said Alexander.

"Oh, you will find out this afternoon," said the Sorcerer. "There are several robot stations on this island, each with its own programming system. After lunch, we will visit the laboratories of some of the engineers in charge. They will explain their systems to you."

I. THE SYSTEM OF CHARLES ROBERTS

After an excellent lunch, cooked and served by skilled robots, the Sorcerer took our two friends to the Northeast Station, whose director was a pleasant individual named Charles Roberts. After introducing his students to Roberts, the Sorcerer took his leave, explaining that he had some things to attend to on the island.

"Now for the details of my system," said Roberts with a smile. "As you have observed, each robot is labeled with a string of letters. These letters constitute a program determining just what the robot will do.

"Let me explain to you my terminology and notation," he continued. "By an *expression* or a *combination*, I mean any string of capital letters—for example, MLBP is an expression; so is LLAZLBA; so is a single letter like G standing alone. I use small letters like x, y, z, a, b, c to stand for particular strings of capital letters, and by xy I mean the combination x followed by the combination y. For example, if x is the expression MBP and y is the expression SLPG, then xy is the expression MBPSLPG.

Also, Ax is then AMBP; xA is MBPA; GxH is GMBPH; CxLy is CMBPLSLPG. Do you get the idea?"

His visitors had no trouble grasping this.

"Now," explained Roberts, "my programming system is based on the idea of certain expressions being *names* of others. And I have two rules concerning the naming of expressions. My first rule is:

Rule Q.   For any expression x, the expression Qx names x.

"Thus, for example, QBAF names BAF; QQH names QH; QDCD names DCD.

"We might refer to Qx as the *principal* name of x, but, as you will see, an expression x may have other names as well. My second naming rule is the following:

Rule R.   If x names y, then Rx names yy.

"Thus, for example, RQB names BB, since QB names B. Or again, RQBR names BRBR, since QBR names BR. In general, for any expression x, the expression RQx names xx, since Qx names x. Do not make the common mistake of believing that Rx names xx; in general it does not. It is RQx that names xx.

"I call Rule R the *repetition* rule, because for any x, the expression xx is called the *repeat* of x—thus, ABCABC is the repeat of ABC. And so Rule R tells us that if x names y, then Rx names the repeat of y.

"To see if you have grasped these rules, what does RRQBH name?"

After a moment's thought, Alexander said: "It names BHBHBHBH."

"Why?" asked Roberts.

"Because RQBH names BHBH, hence RRQBH must name the repeat of BHBH, which is BHBHBHBH."

"Good!" said Roberts.

"You said before," said Alexander, "that an expression can have more than one name. How can that happen?"

"Oh," said Roberts, "for example, xx has RQx as one name and Qxx as another. Both name xx. Or again, Qxxxx names xxxx, and so does RQxx, and so does RRQx. Thus Qxxxx, RQxx, RRQx are three different names for the same expression."

"Oh, of course!" said Alexander.

"How is all this related to some robots creating others?" asked Annabelle.

"Oh, my creation rule is a very simple one," said Roberts. "Remember that when we say that x creates y we mean that any robot with program x will create a robot with program y. Here is my creation rule.

Rule C.   If x names y, then Cx creates y.

"Thus, for any y, robot CQy (that is, any robot with program CQy) creates robot y (a robot with program y).

"It also follows from our three rules that CRQx creates xx (since RQx names xx) and CRRQx creates xxxx (since RRQx names xxxx). And now, for some interesting applications."

• 1 •

"You have seen self-producing robots on this island, robots that create duplicates of themselves?"

"Oh, yes," said Annabelle.

"Well, can you now give me a program for one. That is, can you find an x such that x creates x?"

• 2 •

"Very good," said Roberts, after the couple showed him their solution.

"And now," he continued, "before we turn to any more

problems about creation, there are some basic principles about naming that you should grasp. For one thing, can you find some x that names itself?"

· 3 ·

"Excellent," said Roberts. "Can you find another x that names itself, or is there only one?"

· 4 ·

"You said before," said Alexander, "that in general, Rx does not name xx. Can it ever happen, by some sort of coincidence maybe, that Rx *does* name xx? That is, is there any x such that Rx names xx?"

"What a curious question," said Roberts. "I never thought about that before. Let's see now."

At this point, Roberts took out pencil and paper and made some calculations. "Yes," he said after a while, "there does happen to be such an expression x. Can you find it?"

5 · Some More Curiosities

"If you like curiosities," said Roberts, "it might amuse you to know that for each of the following conditions there is some x that satisfies that condition.

(a) Rx names x.
(b) RQx names QRx.
(c) Rx names Qx.
(d) RRx names QQx.
(e) RQx names RRx.

"But these are only curiosities," added Roberts, "and have no applications that I know of to robot programming. You might have fun working them out sometime at your leisure, if you like puzzles for their own sake, regardless of practical applications."

Some Fixed-Point Principles   "We saw some robots destroying others," said Alexander. "Does your program provide for these?"

"No," replied Roberts, "the destruction programs are in the hands of my brother, Daniel Chauncey Roberts, who is also a roboteer on this island. His programs are very interesting, and you should make a point of visiting him today.

"And now, I will tell you of a basic principle underlying many of my programs as well as those of my brother."

6 • The Fixed-Point Principle

"Given an expression a, we call x a *fixed point* of a if x names ax. (Remember that a, like x and y, represents any given combination of capital letters.)

The fixed-point principle is that every expression a has a fixed point. Moreover, there is a simple recipe whereby, given any expression a, one can find a fixed point of a. Can you prove the fixed-point principle and give me the recipe? For example, what x is such that x names ABCx?"

• 7 •

"The fact that there is a self-creating robot is an application of the fixed-point principle. Can you see why? No? Let me put the matter more precisely. Suppose that we consider another naming system in which Rule Q and Rule R are replaced by other naming rules, and we also have the rule that if x names y, then Cx creates y. If also this other naming system had the fixed-point property—if for every expression a, there were some x that names ax—then there would have to be a self-creating robot. Can you see why?"

• 8 •

"As another application of the fixed-point principle," said Roberts, "there must be some x that names xx. Can you find one? Also, for any expression a, there must be some x that names axax. Can you see why?"

• 9 •

"As a further application, find an x that creates xx."

• 10 •

"Now show that for any expression a, there is some x that creates ax. Give a recipe for finding one, given the expression a. This recipe will have further applications, and I will call it the C Recipe."

11 • The Double Fixed-Point Principle

"As another application of the fixed-point principle, it follows that for any expressions a and b, there are expressions x and y such that x names ay and y names bx. I refer to this as the double fixed-point principle. There are, in fact, two different ways of finding x and y, given a and b. I shall refer to these recipes as *double fixed-point recipes*. What are they?"

• 12 •

"I saw two distinct-looking robots," said Annabelle, "such that each could create the other. Does your program allow for this?"

"Of course," said Roberts, "You should now be able to find such an x and y. What are they?"

• 13 •

"It is also possible," said Roberts, "to find expressions x and y such that each names the other, yet they are distinct. Can you find them?"

• 14 •

"Also it is possible to find x and y such that x creates y and y names x. There are two solutions. Can you find both?"

• 15 •

"There is also a triple fixed-point principle," said Roberts. "Given any three expressions, a, b, and c, there are expressions x, y, and z such that x names ay, y names bz and z names cx. There are three different recipes for finding them. What are they?"

• 16 •

"Now find expressions x, y, and z such that x creates y, y names z, and z creates x. I'll be satisfied with just one solution."

"And now," said Roberts after our couple had solved the last problem, "I think you are ready to visit my brother, Daniel Chauncey."

He then gave them directions to Daniel Chauncey's laboratory, and after thanking him, they went off to visit the other Professor Roberts.

II. THE SYSTEM OF DANIEL CHAUNCEY ROBERTS

"As my brother has told you," said Daniel Chauncey, "I have programs for robots that destroy others. I also have programs for robots that create others.

"I say that x *destroys* y if any robot labeled x destroys any robot labeled y. Now, my creation rules are exactly the same as those of my brother. As for the destruction rules, I need only one, which is the following:

Rule D. If x creates y, then Dx destroys y.

"For example, if x is any expression, then DCQx destroys x, since CQx creates x. Also DCRQx will destroy xx, since CRQx creates xx."

• 17 •

"What happens," asked Annabelle, "if x destroys y, where x and y are the same expression? In other words, suppose x destroys x. Does this mean that robot x will destroy itself?"

"Exactly!" said Daniel Chauncey. "Such robots we call *suicidal* robots. And now, can you find me an x such that robot x is suicidal? My program provides one."

"That is obvious," said Alexander. We already know how to find an x such that x creates x. Then Dx is self-destroying."

"Not so!" said Daniel Chauncey. "If x is self-creating, then

Dx will destroy x, but that doesn't mean that Dx will destroy Dx, which is what we want."

"Oh, of course!" said Alexander.

What is the correct solution?

• 18 •

"Another useful principle," said Daniel Chauncey, "is that for any expression a there is some x that destroys ax. Can you give me a recipe for one? I call this the D Recipe."

• 19 •

"It is also true that for any expression a there is some x that destroys the repeat of ax—that is, x destroys axax. Can you see why?"

• 20 •

"We also saw another sad situation," said Annabelle. "We saw a robot x create a robot y who then ungratefully destroyed its creator, robot x. Does your system provide for such a program?"

"Yes indeed," replied Daniel Chauncey. "You should be able to find an x and some y such that x creates y and y destroys x. In fact, there are two solutions."

What are they?

• 21 •

"We also saw two distinct robots destroy each other," said Alexander. "Does your system provide for this?"

"Certainly," replied Daniel. "There is some x and some y distinct from x such that x and y destroy each other."

What is the solution?

• 22 •

"Something else we saw," said Annabelle. "We saw a robot x create a robot y who created a robot z who then destroyed x."

"That can be done in my system," said Daniel. "Try to find such an x, y, and z." (Actually, there is more than one solution.)

Friends and Enemies.   "Something I observed," said Alexander, "is that several of the robots had labels beginning with F and several with E. Do these two letters have any special significance?"

"Oh, yes indeed!" replied Daniel Chauncey. "You see, our robots have formed various friendships and enmities. If x creates y, then Fx is a friend of y and Ex is an enemy of y. Also, the *best* friend of y is FCQy and the worst enemy of y is ECQy.

"This leads to some interesting results," said Daniel. "I'll give you some examples."

• 23 •

"A robot who is a friend of himself is called *narcissistic*," Daniel continued. "There is an x that is narcissistic. Can you find it?

"Also, there is a program for a robot x who is an enemy of himself. Such a robot is called *neurotic*. If you can find a program for a narcissistic robot, then of course you can find one for a neurotic robot."

· 24 ·

"It is obviously impossible for an x to be his own *best* friend (no x can equal FCQx), but there is an x who *creates* his own best friend. Find one."

· 25 ·

"Now find an *x* that destroys his worst enemy."

"Of course," continued Daniel, "there must also be some x that destroys his best friend—just replace E by F in the solution to the last problem."

"Why would any robot want to do such a horrible thing as destroy his best friend?" asked Annabelle, quite shocked.

"Unfortunately, some of our robots are insane," replied Daniel.

"Then why do you have programs for insane robots?" persisted Annabelle.

"We can't avoid it," replied Daniel. "So far, we have been unable to find an interesting program system that doesn't have some bad side effects. The situation is analogous to the human genetic code. It unfortunately allows for pathologies to arise."

"Tell me this," said Annabelle. "Is it possible for a robot x to be a friend of a robot y who is an enemy of x?"

"Oh yes," said Daniel. "Such robots are called *Christlike*. After all, Jesus said that we should love our enemies. In fact, there is an x such that he is a friend of his *worst* enemy. Such an x is called a *messiah*.

"Also, a robot who is an enemy of someone who is his friend is called *evil*, and if he is an enemy of someone who is his best friend, then he is called *satanic*."

• 26 •

"There are, in fact, two Christlike robots," continued Daniel, "one of whom is even a messiah; and there are two evil robots, one of whom is even satanic. Can you find programs for them?"

• 27 •

"There is also an x which creates some y which destroys the best friend of x. Can you find it?"

• 28 •

"There is also an x who creates the best friend of some y that destroys the worst enemy of x. Can you find one?"

• 29 •

"And then there is some x who is the best friend of someone who destroys x's worst enemy. Find such an x."

"This is only the beginning of the possibilities," said Daniel Chauncey. "The combinations are endless, and that is what makes this island so interesting."

"I'm really intrigued by these programs," said Annabelle, "but one thing puzzles me. I can understand your interest in the software, since it presents problems that are of combinatorial interest. But what is the point in actually *constructing* these robots? Why are you not satisfied with just the programs? Why do you take the trouble to carry them out?"

"There are several reasons," replied Daniel. "For one thing, it's nice to have empirical confirmation that our programs actually work. But more important, it's of great interest to see how all these robots with their various programs interact with each other. Since we have several hundred robots on this island, it is virtually impossible to predict what will happen in the future. Can this colony of robots survive indefinitely, or will the members one day all get destroyed and nothing remain but dismantled parts? All sorts of interesting sociological questions arise, and we are about to engage a robot sociologist to study the sociology of the robot community. It's very possible that this may throw light on the sociology of human communities."

"Well, it all sounds very interesting," said Annabelle, "and I am very much intrigued by your system."

"My system is a modernized version of an older system devised by Professor Quincy," replied Daniel. "He was the first roboteer on this island. He is now retired, but is still actively doing research, and he likes to receive visitors who are interested in seeing his system. Why don't you pay him a visit?"

At this point the Sorcerer entered the laboratory. "Thought I'd find you here," he said, "and I certainly agree it's a good idea to visit Professor Quincy. I'm not familiar with his system, so I'll take you there."

The three thanked Professor Roberts for his time and went off to visit Professor Quincy.

Solutions

1. For any expression x, the expression CRQx creates the repeat of x (since RQx names the repeat of x), and so we take CRQ for x and we have CRQCRQ create the repeat of CRQ, which is CRQCRQ. And so CRQCRQ creates CRQCRQ. Thus CRQCRQ is our solution.

2. RQRQ names the repeat of RQ, which is RQRQ. Thus RQRQ names itself.

3. The only expressions that name anything must be of one of the forms Qy, RQy, RRQy, RRRQy, and so forth. That is, any "name" is either of the form Qy, for some y, or Qy preceded by one or more R's.

We want a "namer" x that names itself. Could it be of the form Qy? Could Qy name Qy? Certainly not; Qy names y and y cannot equal Qy (it is shorter).

What about an expression of the form RQy? Well, RQy names yy, and we want RQy to name RQy, and so this can happen if and only if yy = RQy. Is this possible? Yes, take y = RQ. Thus RQRQ names RQRQ, which is the solution we got before. Moreover, the *only* y such that yy = RQy is RQ. Thus the *only* x of the form RQy that names itself is RQRQ. (Do I hear any R-Q-ment?)

What about an x of the form RRQy? Could that work? No, because RRQy names yyyy, which is necessarily different from RRQy (because if they were the same, we would have to have y = R and y = Q, which is impossible).

RRRQy couldn't work either, because it names yyyyyyyy, which is necessarily longer than RRRQy.

With four or more R's at the beginning, the disparity in lengths would be greater still, and so the only expression that names itself is RQRQ.

4. Try x = RRQRQ.

5. (a) QQ   (d) QQ
   (b) QR   (e) RR
   (c) QQQ

6. RQaRQ names the repeat of aRQ, which is aRQaRQ. And so if we take x = RQaRQ, then x names ax.

Again recall that I am using the small letter a to represent *any* combination of capital letters. Thus, for example, taking

ABC for a, an expression x that names ABCx is RQABCRQ.

Perhaps it is easiest to think of it this way: However one fills in the blanks, the following is true: an x that creates—x is RQ—RQ.

7. In any naming system, if there is some y that names Cy, then Cy creates Cy, and Cy is then a self-creating expression (Robot Cy creates Robot Cy). In this particular naming system, an expression y that names Cy is RQCRQ (by the fixed-point recipe), and thus CRQCRQ is self-creating (which is the solution we originally had).

8. If we can get some y that names Ry, then Ry will name the repeat of Ry, and we can then take x to be Ry. Well, by the fixed-point recipe, y = RQRRQ (this is a special case of the recipe in which a is the letter R). So we take x = RRQRRQ, and the reader can check that it names RRQRRQRRQRRQ.

Also, given any expression a, to find an x that names the repeat of ax, we first look for some y that names aRy—such an expression y is RQaRRQ (by the fixed-point recipe, taking aR for a)—and then Ry names the repeat of aRy, and so we take x = Ry. Thus our solution is x = RRQaRRQ.

9. By the solution of the last problem, taking C for a, there is some y that names CyCy—namely, y = RRQCRRQ. Then Cy must create CyCy, and so we take x = CRRQCRRQ.

10. What we need is some y that names aCy, and then Cy creates aCy, so we then take x = Cy. Using the fixed-point recipe, we take y = RQaCRQ. And so our solution is CRQaCRQ. This is our C Recipe.

11. We want x to name ay and y to name bx. There are two ways we can go about it. One way is to get an x that names aQbx and then take y to be Qbx. (Then, obviously, x names ay and y, which is Qbx, names bx.) We use the fixed-point recipe for getting such an x, and we get the solution

$$x = RQaQbRQ$$
$$y = QbRQaQbRQ$$

Another way is first to get some y that names bQay and take $x = Qay$. This gives the following solution:

$$x = QaRQbQaRQ$$
$$y = RQbQaRQ$$

12. All we need is an x that creates CQx. Then CQx in turn creates x. We use the C Recipe, taking CQ for a, and we get $x = CRQCQCRQ$, which creates the repeat of CQCRQ, which is CQCRQCQCRQ, which is CQx. Thus two distinct expressions that create each other are CRQCQCRQ and CQCRQCQCRQ.

13. RQQRQ names QRQQRQ, which in turn names RQQRQ.

14. One solution can be obtained by taking an x that creates Qx and taking $y = Qx$. This gives the following solution:

$$x = CRQQCRQ$$
$$y = QCRQQCRQ$$

Another solution is obtained by taking some y that names CQy and taking x to equal CQy. This gives the following solution:

$$y = RQCQRQ$$
$$x = CQRQCQRQ$$

15. One solution is to take an x that names aQbQcx, and then take $z = Qcx$ and $y = Qbz$. This gives the following solution:

$$x = RQaQbQcRQ$$
$$z = QcRQaQbQcRQ$$
$$y = QbQcRQaQbQcRQ$$

Another solution is obtained by taking y that names bQcQay and then taking x = Qay and z = Qcx. We leave this to the reader.

Another solution is to take some z that names cQaQbz and then take y = Qbz and x = Qay. We leave this to the reader.

16. One way to go about it is this: Hold x in abeyance for the moment and whatever we finally decide for x, we will take z to be CQx (which creates x). Then y is to name z, so we will take y = Qz. Thus y = QCQx. And so we want x such that it creates QCQx. Using the fixed-point recipe, we get the following solution:

$$x = CRQQCQRQ$$
$$y = QCQCRQQCQRQ$$
$$z = CQCRQQCQRQ$$

There are two other solutions that we leave to the interested reader.

17. DCRQDCRQ.

18. Take x = DCRQaDCRQ.

19. Take y = DCRRQaDCRRQ.

20. One solution is obtained by taking an x that creates DCQx and taking y that equals DCQx. We use the C Recipe (Problem 10) and get the following x and y:

$$x = CRQDCQCRQ$$
$$y = DCQCRQDCQCRQ$$

Another solution is obtained by taking a y that destroys CQy and taking an x that equals CQy. We use the D Recipe (Problem 18) and get the following solution:

$$y = DCRQCQDCRQ$$
$$x = CQDCRQCQDCRQ$$

21.   We want an x that destroys DCQx. We use the D Recipe and get x = DCRQDCQDCRQ. This x destroys DCQDCRQDCQDCRQ, which in turn destroys x.

22.   One solution is obtained by getting an x that creates CQDCQx, which we take for y, and then taking DCQx for z. Using the C Recipe, we get the following solution:

$$x = \text{CRQCQDCQCRQ}$$
$$y = \text{CQDCQCRQCQDCQCRQ}$$
$$z = \text{DCQCRQCQDCQCRQ}$$

23.   FCRQFCRQ is a friend of the repeat of FCRQ which is FCRQFCRQ. Thus FCRQFCRQ is narcissistic.

Obviously, then, ECRQECRQ is neurotic (an enemy of himself).

24.   This is obvious: We want x to create FCQx, so by the C Recipe we take x = CRQFCQCRQ.

25.   Take x = DCRQECQDCRQ.

26.   A messiah is an x who is a friend of ECQx (the worst enemy of x). So x must be of the form Fz for some z that creates ECQFz (which is ECQx). We use the C Recipe and take z = CRQECQFCRQ. Then we take x = FCRQECQFCRQ and x is our messiah. Now, z creates ECQFz, which is ECQFCRQECQFCRQ, which we will call y. Although x is a friend of y, it is not the best friend of y, and so y is evil (being an enemy—in fact, the worst enemy—of x), but not satanic.

Of course, by interchanging E and F, we get a satanic x and a Christlike y who is not a messiah. In summary:

FCRQECQFCRQ is a messiah.

ECQFCRQECQFCRQ is evil but not satanic.

ECRQFCQECRQ is satanic.

FCQECRQFCQECRQ is Christlike but not a messiah.

27. We will take an x that creates DCQFCQx. Such an x is CRQDCQFCQCRQ.

28. We will take an x that creates FCQDCQECQx. Let y = DCQECQx. Then x creates FCQy, the best friend of y, and y destroys ECQx, the worst enemy of x. And so we get the solution x = CRQFCQDCQECQCRQ.

29. We would like x and y to be such that x = FCQy (the best friend of y) and y destroys ECQx. Thus we want y to destroy ECQFCQy. We thus take y = DCRQECQFCQDCRQ, and so our desired x is FCQDCRQECQFCQDCRQ.

12

THE QUAINT SYSTEM OF PROFESSOR QUINCY

I. QUINCY'S SYSTEM

"AH YES," said Quincy, after his three guests were comfortably seated, "my system may be old-fashioned, but it really works! Yes, it really works, you know," he repeated with a chuckle.

"My system is based on the idea of *quotation*—you know, one names an expression by enclosing it in quotation marks, the opening quote on the left and the closing quote on the right. Only instead of using opening and closing quotes, I use the symbol Q_1 for the opening quote and Q_2 for the closing quote. And so by the *quotation* of an expression x, I mean the expression $Q_1 x Q_2$—for example, the quotation of HFUG is $Q_1 HFUG Q_2$. My first rule is the following:

Rule $Q_1 Q_2$. $Q_1 x Q_2$ names x.

"Thus the *quotation* of x names x. Compare this with the first naming rule of Roberts's system—Qx names x. His system is based on the more modern idea of *one-sided* quotation, which today is used in many computer programming languages such as LISP. Such systems use only an opening quote, which for Roberts is the symbol Q.

"Next, by the *norm* of an expression, I mean the expression followed by its own quotation. Thus, the norm of x is xQ_1xQ_2. This notion of *norm* I have borrowed from the logician Raymond Smullyan. The operation of taking the norm of an expression plays the same fundamental role in my system as repetition does in the Roberts system. And so, instead of Roberts's Rule R, my basic naming rule is:

Rule N. If x names y, then Nx names the norm of y.

"Thus, if x names y, then Nx names yQ_1yQ_2. In particular, NQ_1yQ_2 names yQ_1yQ_2. I have other naming rules too, but first I would like to show you some things that can be done with just these two."

Professor Quincy then gave his three guests the following problems:

• 1 •

Find an x that names itself.

• 2 •

The first fixed-point principle of Roberts's system also holds for the present system—that is, for any expression a, there is some x that names ax. Give a recipe for finding such an x, given a.

• 3 •

There is an expression that names its own norm, and another that names the norm of its norm. Do you see how to find them?

• 4 •

"My creation rule is the same as in the Roberts system," said Quincy. "If x names y (in my system) then Cx creates y. Also my destruction rule is the same as Rule D of the Roberts system—if x creates y, then Dx destroys y.

"Now find an x that creates itself and an x that destroys itself."

• 5 •

"Also, given any expression a, there is an x that creates ax and an x that destroys ax. Can you see why?"

• 6 •

"As with the Daniel Roberts system, if x creates y, then Fx is a friend of y and Ex is an enemy of y. But in my system, the best friend of y is FCQ_1yQ_2, and the worst enemy of y is ECQ_1yQ_2.

"Now find an x that is a friend of x and an x that is an enemy of x."

"In the Roberts system," said Annabelle, "one can find an x that creates a friend of x. Can this be done in your system?"

"Probably not, with the rules I have given you so far. There are two more naming rules to come; the first is:

Rule M. If x names y then Mx names Q_1yQ_2 (the quotation of y).

"With this rule, one can do several things that apparently one could not do before. I will give you some examples."

• 7 •

Find an x that names its own quotation.

• 8 •

Find an x that names the norm of its own quotation, and another x that names the quotation of its norm.

• 9 •

Show that for any expression a there is some x that names the quotation of ax.

• 10 •

Show that for any expression a there are expressions x and y such that x names y and y names ax.

"Is it also true," asked Annabelle, "that for any expressions a and b, there are expressions x and y such that x names ay and y names bx, as in the Roberts system?"

"I doubt it," said Quincy, "and that's why I need my final rule, which I am about to give you.

"For any expression x of two or more letters, by $x^{\#}$ I shall mean the result of erasing the first letter of x—for example, if x is BFGH, then $x^{\#}$ is FGH. If x consists of only one letter, then we will take $x^{\#}$ to be x itself (this is for 'waste' cases). My final rule, then, is the following erasure rule:

Rule K.   If x names y, then Kx names $y^{\#}$.

"For example, KQ_1BFGHQ_2 names FGH (since Q_1BFGHQ_2 names BFGH).

"With this rule added, my programming system is complete, and except for a few curiosities, we can do just about everything that can be done in the Roberts system. I'll give you some examples."

Quincy then presented the problems that follow. The first one is really the basis for the others.

11 • The Quincy Recipes

Given any expression a, one can find an x that names axQ_2, also an x that names axQ_2Q_2, also an x that names $axQ_2Q_2Q_2$, and the same for any number of Q_2's.

· 12 ·

Now one can establish the double fixed-point principle: For any a and b there are x and y such that x names ay and y names bx. And there are two recipes for finding x and y. What are they? And what about the triple fixed-point principle?

13 · More Quincy Recipes

Given a, it is useful to have recipes on hand to find an x that

(1) creates axQ_2

(2) creates axQ_2Q_2

(3) creates $axQ_2Q_2Q_2$

These will be used in subsequent problems; we shall call them Quincy C Recipes.

It is also useful to have on hand a recipe for finding an x that

(4) destroys axQ_2

This, however, can be got from (1) by replacing C with DC; we call this a Quincy D Recipe.

Now, find these recipes.

· 14 ·

Find an x and y such that x creates y and y destroys x. (There are two solutions.)

· 15 ·

Find an x that creates its best friend and an x that destroys its worst enemy.

· 16 ·

Find an x that is a messiah (a friend of its worst enemy) and a satanic x (an enemy of its best friend).

• 17 •

Find an x who is the best friend of someone who destroys x's worst enemy.

• 18 •

Find an x who creates the best friend of someone who destroys x's worst enemy.

"This should give you an idea of how my system works," said Quincy. "The possibilities are really endless."

II. CAN YOU FIND THE KEY?

"If you like curiosities and complex problems," said Quincy, "I must tell you of a curious system devised by Professor Cudworth, who is another roboteer on this island.

"He was curious to know what would happen with my system if the normalization rule—Rule N, which is really at the heart of my system—was replaced by the repetition rule—Rule R of Roberts. He tried this and got nowhere. Yes, he could solve a few trivial problems, but they were useless for constructing programs. Indeed, he couldn't even get a program for a self-duplicating robot—that is, he couldn't even get an x such that x creates x. Then, a friend of his, Inspector Craig of Scotland Yard, who takes a great interest in combinatorial puzzles suggested that Cudworth add the reversal rule, and when he did so, the system worked fine! It could do everything that could be done in my system, and also everything that could be done in Roberts's system, and more!"

"What is the reversal rule?" asked Alexander.

"Oh, by the *reverse* of an expression is meant the expression written backwards. For example, the reverse of ABCD is

DCBA. Well, Cudworth took the letter V and added the rule that if x names y, then Vx names the reverse of y. Then, as I said, the system worked perfectly."

Quincy then reviewed the rules of Cudworth's system for his three guests:

Rule Q_1Q_2. Q_1xQ_2 names x.

Rule R.   If x names y, then Rx names yy.

Rule M.   If x names y, then Mx names Q_1yQ_2.

Rule K.   If x names y, then Kx names $y^\#$.

Rule V.   If x names y, then Vx names the reverse of y.

"The creation, destruction, friendship and enmity rules are the same as in my system," continued Quincy. "Now, the first problem is to find an x that names itself. This is not simple! The shortest one I know has eighteen letters. There is also an x that names its own norm. The shortest one that I know has thirty-four letters.

"There is also an x that names its own reverse, one that creates its own reverse, one that destroys the reverse of its repeat, one that is a friend of an enemy of the reverse of the norm of its own quotation, and I could go on endlessly mentioning other combinations. The system also obeys the fixed-point principle, the double fixed-point principle, and so forth.

"You might have fun working these problems out at your leisure. I believe that once you solve the first problem, you will have the key to all the others."

• 19 •

Professor Quincy was right in saying that the essential idea behind the construction of an x that names itself is really the key to solving all the others. What is the key?

When our three friends left Professor Quincy, they felt that it was too late to set sail for home that evening, so they decided to stay overnight at an inn.

"Tomorrow," said the Sorcerer, "before we leave this island, I would like us to visit the station of Simon Simpson, who is the youngest and most progressive roboteer on this island. I've heard he is a bit on the vain side, but also that his program system is a very simple and elegant one, and definitely worth learning."

The three stayed up about half the night trying to decipher Cudworth's strange system, but they finally managed to solve it. Exhausted, they went to sleep and did not rise till late next morning. After breakfast they went off to visit Simon Simpson. What happened will be related in the next chapter.

Solutions

1. NQ_1NQ_2 names the norm of N, which is NQ_1NQ_2. And so NQ_1NQ_2 names itself.

2. Take $x = NQ_1aNQ_2$. It names the norm of aN, which is aNQ_1aNQ_2, which is ax.

3. If y names Ny, then Ny will name the norm of Ny. Then we take $x = Ny$, and x will then name its own norm. According to the solution to Problem 2 (which is the fixed-point recipe for Quincy's system) we can take $y = NQ_1NNQ_2$ (that is, we take N for a), and so an x that names its own norm is NNQ_1NNQ_2.

An x that names the norm of its norm is $NNNQ_1NNNQ_2$.

4. CNQ_1CNQ_2 creates itself (since NQ_1CNQ_2 names CNQ_1CNQ_2).

$DCNQ_1DCNQ_2$ destroys itself.

5.   An x that creates ax is found by first finding some y that names aCy, and then taking x to equal Cy. We take y = NQ_1aCNQ_2, and so our solution is x = CNQ_1aCNQ_2.

An x that destroys ax is $DCNQ_1aDCNQ_2$.

6.   We need some y that creates Fy, and then Fy will be a friend of Fy. We take x = Fy, and x is then a friend of itself. By the last problem (using F for a), we take y = CNQ_1FCNQ_2, and so $FCNQ_1FCNQ_2$ is a friend of itself.

An enemy of itself is $ECNQ_1ECNQ_2$.

Discussion.   Before giving further solutions, let us pause and note that so far these solutions are not very different from those of the Roberts system—just replace N by R and both Q_1 and Q_2 by Q and the above solutions become solutions in the Roberts system. For example, compare NQ_1NQ_2, which names itself in Quincy's system, to RQRQ, which names itself in the Roberts system. Or again, in the last problem, $ECNQ_1ECNQ_2$ is an enemy of itself in Quincy's system, whereas ECRQECRQ is an enemy of itself in the Roberts system. But in the problems that follow, the situation changes quite noticeably.

7. MNQ_1MNQ_2 works, because NQ_1MNQ_2 names the norm of MN, which is MNQ_1MNQ_2. Hence MNQ_1MNQ_2 names the quotation of MNQ_1MNQ_2.

8.   An x that names the norm of its own quotation is $NMNQ_1NMNQ_2$. An x that names the quotation of its norm is $MNNQ_1MNNQ_2$.

9.   Take x = MNQ_1aMNQ_2.

10.   This follows from the last problem: Since there is some x that names the quotation of ax, we can take y to be the quotation of ax, and so y then names ax. Hence the solution is x = MNQ_1aMNQ_2; y = $Q_1aMNQ_1aMNQ_2Q_2$.

We note that with Quincy's final rule (Rule K) there is a second solution that we will discuss later.

11. An x that names axQ_2 is $KMNQ_1aKMNQ_2$, because NQ_1aKMNQ_2 names $aKMNQ_1aKMNQ_2$, which is ax, hence MNQ_1aKMNQ_2 names Q_1axQ_2, hence $KMNQ_1aKMNQ_2$ names axQ_2.

Remark. This gives a second solution to the last problem: We can take y such that it names aQ_1yQ_2 (by taking aQ_1 instead of a)—namely, $y = KMNQ_1aQ_1KMNQ_2$. Then, we take x to be the quotation of y—i.e., $x = Q_1KMNQ_1aQ_1KMNQ_2Q_2$.

An x that names axQ_2Q_2 is $KKMMNQ_1aKKMMNQ_2$. And an x that names $axQ_2Q_2Q_2$ is $KKKMMMNQ_1a$ $KKKMMMNQ_2$, and so forth.

12. One way is to get an x that names aQ_1bxQ_2 and then take $y = Q_1bxQ_2$. We use the Quincy recipe (taking aQ_1b in place of a) and get $x = KMNQ_1aQ_1bKMNQ_2$. We then take $y = Q_1bKMNQ_1aQ_1bKMNQ_2Q_2$. (The reader can verify directly that this x and y works.)

Another solution is obtained by first taking y that names bQ_1ayQ_2 and then taking $x = Q_1ayQ_2$. We then get the solution $x = Q_1aKMNQ_1bQ_1aKMNQ_2Q_2$ and $y = KMNQ_1b$ Q_1aKMNQ_2.

As for 3-cycles, we can take x such that it names $aQ_1bQ_1cxQ_2Q_2$ (namely, $x = KKMMNQ_1aQ_1b$-$Q_1cKKMMNQ_2$) and then take $z = Q_1cxQ_2$ and $y = Q_1bzQ_2$. We leave it to the reader to write out z and y explicitly.

Another solution can be obtained by taking y such that it names $bQ_1cQ_1ayQ_2Q_2$, then taking $x = Q_1ayQ_2$ and $z = Q_1cyQ_2$. Another solution can be obtained by taking z that names $cQ_1aQ_1bzQ_2Q_2$, and taking $y = Q_1bzQ_2$ and $x =$

Q_1ayQ_2. These solutions can be written out explicitly, if so desired.

13. To find an x that creates axQ_2, we need some y that names $aCyQ_2$ and then take $x = Cy$. We use the first Quincy recipe of Problem 10 and get $y = KMNQ_1aCKMNQ_2$. And so our desired x is $CKMNQ_1aCKMNQ_2$.

An x that creates axQ_2Q_2 is $CKKMMNQ_1aCKKMMNQ_2$.

An x that creates $axQ_2Q_2Q_2$ is $CKKKMMMNQ_1a$-$CKKKMMMNQ_2$.

An x that destroys axQ_2 is $DCKMNQ_1aDCKMNQ_2$.

14. One way is to take an x that creates DCQ_1xQ_2, which in turn destroys x. Using the Quincy C Recipe, we get the solution

$$x = CKMNQ_1DCQ_1CKMNQ_2$$
$$y = DCQ_1CKMNQ_1DCQ_1CKMNQ_2Q_2$$

Another way is to take some y that destroys CQ_1yQ_2 and take $x = CQ_1yQ_2$. We then get

$$x = CQ_1DCKMNQ_1CQ_1DCKMNQ_2Q_2$$
$$y = DCKMNQ_1CQ_1DCKMNQ_2$$

15. This is now simple: By the Quincy C Recipe, an x that creates its best friend FCQ_1xQ_2 is $CKMNQ_1FCQ_1CKMNQ_2$.

An x that destroys its worst enemy is $DCKMNQ_1$-$EDCQ_1DCKMNQ_2$.

16. To get a messiah, we need some z that creates ECQ_1FzQ_2 (the worst enemy of Fz) and Fz is then a messiah. Using the Quincy C Recipe, we take $z = CKMNQ_1ECQ_1FCKMNQ_2$. And so, our messiah is $FCKMNQ_1ECQ_1FCKMNQ_2$.

A satanic x is $ECKMNQ_1FCQ_1ECKMNQ_2$.

17. Let y be this "someone" who destroys x's worst enemy. Then x is the best friend of y, so $x = FCQ_1yQ_2$. The worst enemy of x is ECQ_1xQ_2, which in terms of y is ECQ_1F-$CQ_1yQ_2Q_2$. And so we need y to create $ECQ_1FCQ_1yQ_2Q_2$. We use a Quincy C Recipe and get y = $CKKMMNQ_1$-$ECQ_1FCQ_1CKKMMNQ_2$. And so our x is FCQ_1CKKM-$MNQ_1ECQ_1FCQ_1CKKMMNQ_2Q_2$.

18. Again we let y be the someone who destroys ECQ_1xQ_2 (the worst enemy of x). We will take y to be $DCQ_1ECQ_1xQ_2Q_2$. Then the best friend of y is $FCQ_1DCQ_1ECQ_1xQ_2Q_2Q_2$. And so we need an x that creates $FCQ_1DCQ_1ECQ_1xQ_2Q_2Q_2$. We use a Quincy C Recipe and we get x = $CKKKMMM$-$NQ_1FCQ_1DCQ_1ECQ_1CKKKMMMNQ_2$.

19. The key to all the problems that arise in Cudworth's system is to find an expression s that does the same job as the letter N does in Quincy's system—that is, we want s to be such that for any expressions x and y, if x names y, then sx will name the norm of y. Such an expression s might aptly be called a *normalizer*.

I will use the notation $\bar{y}$ to mean the *reverse* of y. So, how can we get from an expression y to its norm yQ_1yQ_2 by a sequence of operations all of which can be programmed in this system? One such sequence is the following:
(1) Take the quotation of y, thus getting Q_1yQ_2.
(2) Reverse what you now have, getting $Q_2\bar{y}Q_1$.
(3) Erase the Q_2, getting $\bar{y}Q_1$.
(4) Reverse what you have, getting Q_1y.
(5) Repeat what you now have, getting Q_1yQ_1y.
(6) Erase the leftmost Q_1, getting yQ_1y.
(7) Now take its quotation, getting $Q_1yQ_1yQ_2$.
(8) Erase the leftmost Q_1, getting yQ_1yQ_2.

And so suppose x names y. Then
(1) Mx names Q_1yQ_2.
(2) VMx names $Q_2\bar{y}Q_1$.

(3) KVMx names $\overline{y}Q_1$.

(4) VKVMx names Q_1y.

(5) RVKVMx names Q_1yQ_1y.

(6) KRVKVMx names yQ_1y.

(7) MKRVKVMx names $Q_1yQ_1yQ_2$.

(8) KMKRVKVMx names yQ_1yQ_2.

And so, KMKRVKVM is a normalizer. All the problems of this chapter can be solved in Cudworth's system by taking KMKRVKVM in place of the single letter N.

For example, an expression that names itself is KMKRVK-VMQ$_1$KMKRVKVMQ$_2$. (Is there a shorter one?) An expression that creates itself is CKMKRVKVMQ$_1$CKMKRVKVMQ$_2$.

I will leave to the reader, as an exercise, the construction of an expression that names its own reverse. Also, what about an expression that creates its own reverse? And what about one that creates the repeat of its reverse?

13

FROM THE RIDICULOUS TO THE SIMPLE

"THOSE OLD-FASHIONED SYSTEMS really get me," said Simon Simpson, shaking his head. "Do you know, for example, the weird system of Cudworth?"

"Yes," said the Sorcerer, "Professor Quincy showed it to us yesterday."

"Now, if that isn't the craziest system I've ever seen in my life! To go to all that trouble just to get a normalizer! Sometimes I wonder about the sanity of some of my colleagues!"

"Oh, I don't know," said the Sorcerer. "I found it quite an interesting challenge to find the key to the system."

"But the system is so unnecessarily complicated!" said Simpson. "Cudworth seems to delight in making things as difficult as possible. My philosophy is the very opposite—I like to make things as simple as possible.

"Actually, Quincy's system is not all that bad," Simpson went on, "and considering that it was devised before one-sided quotation was discovered, it is understandable. Still, the last rule—the erasure rule—seems artificial and nothing more than an ad hoc device to remedy a none-too-good situation. But even here, Quincy could have done better: he could have replaced the two rules—Rule M and Rule K—by a single rule: if x names

141

y, then Lx names yQ_2. That one rule would have yielded the double and triple fixed-point principles, and all the problems that he probably gave you would be solvable.

"Of course, the Roberts system is much cleaner and more natural. But even this system is more complex than it need be for the problems that really matter from a practical point of view. I am not interested in such academic questions as whether an x can be found that creates its own repeat; of what significance is that to robotry? My approach is purely pragmatic. I'm interested only in questions of sociological importance: Which robots create which? Which destroy which? Which are friends, best friends, enemies, worst enemies, of which? And for matters like these, my program system is the most efficient of all."

"Do you use one-sided or two-sided quotation?" asked the Sorcerer.

"Neither; my system is quotationless. I don't bother with certain expressions *naming* others."

"That's interesting," said the Sorcerer. "I have also been experimenting of late with quotationless systems—not for robotry, but for certain general problems that arise with self-reference. I am really curious to hear about your system."

"My rules are direct, short, sweet, and to the point. I use the symbols C, D, F, E, Ċ, Ḋ, Ḟ, Ė and my rules are these:

Rule C.   Cx creates x.

Rule Ċ.   Ċx creates xx.

Rule D.   Dx destroys x.

Rule Ḋ.   Ḋx destroys xx.

Rule F.   Fx is the best friend of x.

Rule Ḟ.   Ḟx is a friend of xx.

Rule E.   Ex is the worst enemy of x.

Rule Ė.   Ėx is an enemy of xx.

"What could be more direct than these? The solutions to all the problems I will now give you are shorter than any you have yet seen. For example, it is obvious that a self-reproducing robot is ĊĊ and a self-destroying one is ḊḊ. So here are some problems, and I am sure you will solve them quite easily. And incidentally, the solutions place the solutions in Roberts's system in a clearer light, in a manner I will later explain. But for now, let's concentrate on just the problems." Professor Simpson's problems follow.

• 1 •

Find a distinct x and y such that each creates the other.

• 2 •

Find an x and y such that x creates y and y destroys x. There are two solutions.

• 3 •

Show that for any expression a there is some x that creates ax, and some x that destroys ax.

• 4 •

Given any expressions a and b, show:

(a) There are an x and y such that x creates ay and y creates bx. (There are two solutions.)

(b) There are an x and y such that x destroys ay and y destroys bx. (Two solutions.)

(c) There are an x and y such that x creates ay and y destroys bx.

• 5 •

Find an x that is a friend of itself.

• 6 •

Find an x that creates its best friend.

• 7 •

Find an x that creates a friend that is not its best friend.

• 8 •

Find an x that is a friend of its worst enemy.

• 9 •

Find an x that is the best friend of one of its enemies.

• 10 •

Find an x and y such that x creates the best friend of y and y destroys the worst enemy of x.

• 11 •

Find an x that is the best friend of one that destroys its worst enemy.

• 12 •

Find an x that creates some y that is a friend of some z that is the worst enemy of some w that destroys the best friend of the worst enemy of x.

"And so you see," said Simpson proudly, "that all sorts of complicated sociological situations can be programmed very easily in my system."

"Your system is indeed neat and economical," said the Sorcerer, "and I like it very much. It has many similarities to a system of mine, and by a strange coincidence you use the dot over a letter in much the way I do.

"One thing you may not realize is that all your solutions can be easily transformed into solutions in the Roberts system simply by replacing C by CQ, Ċ by CRQ, D by DCQ, Ḋ by DCRQ, F by FCQ, Ḟ by FCRQ, E by ECQ, and Ė by ERCQ. For example, your x that creates itself is ĊĊ. If we replace Ċ by CRQ, we get CRQCRQ, which is the x of the Roberts system that creates itself. The same holds for all your solutions."

"I certainly *do* realize that," said Simpson, "and that is what I meant before when I said that my solutions put the Roberts solutions in a clearer light."

The reader can easily check that in each of the twelve problems above, the solutions in the Simpson system can be transformed into solutions in the Roberts system by just the method stated by the Sorcerer.

Solutions

1. ĊCĊ and CĊCĊ

2. *Solution 1:* x = ĊDĊ, y = DĊDĊ
   Solution 2: x = CḊCḊ, y = ḊCḊ

3. An x that creates ax is ĊaĊ. An x that destroys ax is ḊaḊ.

4. (a) *Solution 1:* x = ĊaCbĊ, y = CbĊaCbĊ
   Solution 2: x = CaĊbCaĊ, y = ĊbCaĊ
   (b) Same, using Ḋ in place of Ċ and D in place of C
   (c) *Solution 1:* x = ĊaDbĊ, y = DbĊaDbĊ
   Solution 2: x = CaḊbCaḊ, y = ḊbCaḊ

5. ḞḞ

6. ĊFĊ

7. ĊFCḞ

8. ḞEḞ

9. ḞEFĖ (Robot ĖFĖ is satanic)

10. *One Solution:* x = ĊFDEĊ, y = DEĊFDEĊ
    Another Solution: x = CFḊECFḊ, y = ḊECFḊ

11. x = FḊEFḊ

12. There are several solutions. One of them is x = ĊFED-
FEĊ (which creates FEDFEx). Another is x = ĊFEDFECḞ.
Another is CFEḊFECFEḊ.
   In each solution, the y, z, and w can easily be found.

PART IV

GÖDELIAN PUZZLES

14

SELF-REFERENCE AND
CROSS-REFERENCE

A FEW DAYS AFTER their return from the Island of Robots, our couple paid another visit to the Sorcerer.

"I was really intrigued by those programs," said Annabelle. "How did those robot engineers ever think them up?"

"They are all related to problems in designation," replied the Sorcerer, "particularly to quotational designation."

"Just what is that?"

"I'll have to explain it to you from the beginning," said the Sorcerer. "Do you know the difference between the *use* and *mention* of words?"

Neither Annabelle nor Alexander had ever heard of this.

"I'll illustrate with an example," said the Sorcerer, and he then wrote down the following sentence.

(1) ICE IS FROZEN WATER.

"Is that sentence true or false?"

"Obviously true," they both replied.

"All right. Now, what about the following sentence?"

(2) ICE HAS THREE LETTERS.

"Also true," said Annabelle.

"Of course it's true," said Alexander. "Ice *does* have three letters."

"No it doesn't!" said the Sorcerer. "Frozen water doesn't have any letters at all! It's the *word* 'ICE' that has three letters. So sentence (2) as it stands is simply wrong! The correct version is the following.

(2)   "ICE" HAS THREE LETTERS.

"Sentence (1) talks about ice the *substance;* Sentence (2) talks about ice the substance, but what it says is wrong. Sentence (2)' talks about the word 'ICE' and what it says is true. One talks about the word by enclosing it in quotes—at least that's the older way that we will work with for a while. Anyway, in Sentence (1), the word 'ICE' is *used,* since it talks about the substance, not about the word. But in Sentence (2), the word 'ICE' is *mentioned,* or talked *about,* but not used, since the sentence talks about the word, not about ice the substance."

"That seems clear enough," said Annabelle.

"What *is* used in Sentence (2)," continued the Sorcerer, "is not the word 'ICE' but the *name* or *quotation* of that word. It is used to talk about the word. I know you grasp the general idea, yet beginners often lapse into confusing use with mention. Here, let me try the following." The Sorcerer then wrote down the following and asked the couple whether it was true or false.

(3)   " " "ICE" " " HAS THREE PAIRS OF QUOTES.

"That's obviously true," said Annabelle.

"Of course it is!" said Alexander.

"I'm sorry, but you are both wrong," said the Sorcerer. "Yes, what I *wrote down* does indeed have three pairs of quotes, but what it talks *about* has only two pairs of quotes. The correct rendition is the following:

(3)′ " " "ICE" " " HAS TWO PAIRS OF QUOTES.

"It's a bit confusing," said Annabelle.
"All right, maybe this will help. Isn't the following true?"

(4) ICE HAS NO PAIRS OF QUOTES.

"Yes," said Alexander, "the substance ice has no pairs of quotes."
"Then what about the following?" asked the Sorcerer.

(5) "ICE" HAS NO PAIRS OF QUOTES.

"That's false," said one of the two. "I see *a* pair of quotes."
"That's what you *see*," said the Sorcerer, "but what the sentence talks about is the word 'ICE' without any quotes around it. And so sentence (5) is true! What is *used* has one pair of quotes, but what is talked *about*, or *mentioned*, has none."
"I think I begin to see," said Annabelle.
"Good, then consider the following sentence."

(6) " "ICE" " HAS ONE PAIR OF QUOTES.

"I see now that that is true," said Alexander.
"Good. And what about the following?"

(7) " " "ICE" " " HAS TWO PAIRS OF QUOTES.

"Yes, it is true," said Annabelle.
"Good," said the Sorcerer. "And now I would like to illus-

trate the difference between 'use' and 'mention' in an amusing way. Is the following sentence true or false?"

(8)   IT TAKES LONGER TO READ THE BIBLE THAN TO READ "THE BIBLE."

"It certainly *is* true!" laughed Annabelle. "It takes *much* longer!"

"That's a case," said the Sorcerer, "in which 'the Bible' is both used and mentioned in the same sentence.

"Now, let us try this one."

(9)   THIS SENTENCE IS LONGER THAN "THIS SENTENCE."

"Also true," they both said.

"Good. Now tell me in what language is the following sentence. Is it French or English?"

(10)   "LE DIABLE" IS THE NAME OF LE DIABLE.

"I would say both," said Annabelle. "It contains both English and French words."

"That is right," said the Sorcerer. "Now what about the following?"

(11)   "LE DIABLE" IS THE FRENCH NAME OF THE DEVIL.

"Also both," said Alexander. "It also contains both French and English words."

"Wrong this time," said the Sorcerer. "The French words in it are both in quotes, hence talked about but not used. The point is that any English-reading person, even if he did not know a single word of French, could understand the sentence perfectly (and also learn a tiny bit of French in the bargain). On the other hand, the same person could not understand Sentence (10), since he would not know what it is that 'le diable' is the name of."

SELF-REFERENTIAL SENTENCES

"And now," said the Sorcerer, "I would like to get to the more interesting problems of constructing sentences that refer to themselves.

"Suppose we wish to construct a sentence that ascribes to itself a certain property. To be exact, let's take the property of being read by someone named John. How do we construct a sentence that asserts that John is reading that very sentence? Of course, one obvious way to do so is to construct the following sentence.

(12)   JOHN IS READING THIS SENTENCE.

"Obviously, sentence (12) is true only if John is reading Sentence (12). However, that sentence contains the *indexical* word 'this,' and what we want is to accomplish the same self-reference without the use of an indexical."

"What do you mean by an indexical?" asked Annabelle.

"Oh, an *indexical* is a term whose designation depends on its context. For example, 'James Smith' is not an indexical, for in any context it denotes the person James Smith, whereas the word 'I' is typically an indexical because when used by one person it denotes a different person than when used by another. When James Smith says 'I,' he means James Smith, whereas when Paul Jones says 'I,' he means Paul Jones. Another indexical is the word 'you,' whose designation depends on the person to whom it is addressed. Another indexical is the word 'now,' which designates different moments of time when uttered at different points in time.

"The logician Raymond Smullyan has half-humorously given the name *chameleonic terms* to indexicals, which he wrote about in a paper called 'Chameleonic Languages.' As he says in this paper: 'Like a chameleon whose color depends on its surroundings, these words change their denotation from context to context.' A friend of his with a good sense of humor, who had heard of Smullyan's paper before it was published, wrote to him: 'I have heard of your chameleonic languages. I do know what they are, except that I assume they are not what they appear to be.'

"At any rate, I think you now understand what chameleonic or indexical terms are. The phrase 'this sentence' is obviously an indexical; its denotation depends on the sentence in which it appears—providing, of course, that it is *used* and not *mentioned* in the sentence."

"I don't think I follow that," said Annabelle.

"Well, consider the following sentence, and tell me whether it is true or false."

(13)   THIS SENTENCE HAS FIVE WORDS.

"That sentence is true," said Alexander. "It does have five words."

"Right." said the Sorcerer. "Now what about the following?"

(14)   THIS SENTENCE HAS EXACTLY TWO WORDS.

"That's obviously false," said Annabelle.

"Right. Now what about the following?"

(15) "THIS SENTENCE" HAS EXACTLY TWO WORDS.

"Oh, I begin to see what you mean," said Annabelle. "The last thing you wrote is indeed true. Sentence (15) doesn't say that Sentence (15) has exactly two words, which it obviously hasn't, but that the phrase 'this sentence' has exactly two words, which is indeed true. On the other hand, Sentence (14) doesn't say that 'this sentence' has exactly two words, but that the whole of Sentence (14) has exactly two words, which is clearly false—it has six words."

"Very good!" said the Sorcerer. "I see you are becoming able to distinguish between use and mention. You now realize that the phrase 'this sentence' when surrounded by quotes is *not* indexical; it always denotes the two words that appear within the quotes.

"And now I want to explain how self-reference can be obtained *without* using indexicals."

"Why would one want to?" asked Annabelle. "What is wrong with indexicals? They seem very useful!"

"Of course they are," said the Sorcerer. "They work fine for languages like ordinary English that have indexicals, but for formal mathematical systems of the type studied by Kurt Gödel—which is what I'm leading up to—there are no indexicals; hence Gödel had to achieve self-reference without the use of indexicals."

"How did he do that?" asked Alexander.

"That's what I'm coming to. You know that in algebra one uses the letters x and y as standing for unknown numbers, and that on the island we just visited, the robot engineers used these letters as standing for arbitrary expressions of their programming languages. Well, I will now use the letter x as standing for unknown expressions of ordinary English.

"And now, using an essential idea of Gödel's, I will define the *diagonalization* of an expression to be the result of replacing the symbol x by the quotation of the entire expression. For example, consider the following expression:

(1) JOHN IS READING x.

"Expression (1) is not a sentence; it is neither true nor false as it stands, since we don't know what the symbol x stands for. If we replace x by the name of some expression, then (1) becomes an actual sentence, true or false as the case may be. We could indeed replace x by the quotation of Expression (1), thus getting its diagonalization which is:

(2) JOHN IS READING "JOHN IS READING x."

"Now, (2) is an actual sentence; it asserts that John is reading Expression (1), and it is true if and only if John is reading (1). However, (2) is not self-referential; it does not assert that John is reading (2), but that John is reading (1). To achieve a self-referential sentence, instead of starting with Expression (1), we start with the following.

(3) JOHN IS READING THE DIAGONALIZATION OF x.

"Now let's see what the diagonalization of Expression (3) looks like.

(4) JOHN IS READING THE DIAGONALIZATION OF "JOHN IS READING THE DIAGONALIZATION OF x."

"At first, (4) may not appear to make much sense, but a little thought shows that it makes a good deal of sense and, moreover, reveals something very interesting! Sentence (4) says that John is reading the diagonalization of (3), but the diagonalization of (3) is (4) itself. And so Sentence (4) asserts that John is reading the very Sentence (4)! Thus (4) is a self-referential sentence. Credit for this basic idea goes to Gödel.

"I believe it is easier to see this symbolically. Let us use the letter J as an abbreviation of "John is reading," and let us use D to abbreviate "the diagonalization of." Thus (1) in abbreviated notation is this:

(5)   Jx

"Its diagonalization is:

(6)   J "Jx"

"What (6) says is that John is reading the two-letter expression 'Jx'—thus (6) says that John is reading (5). It is not self-referential; (6) doesn't say that John is reading (6). Now consider the following.

(7)   JDx

"Its diagonalization is:

(8)   JD "JDx"

"What (8) says is that John is reading the diagonalization of (7), but the diagonalization of (7) is (8) itself. Thus (8) is self-referential—Sentence (8) asserts that John is reading Sentence (8)!

"Self-reference comes up in a crucial way in Gödel's famous incompleteness theorem, which I will later tell you about. And the idea of diagonalization comes very close to the technique used by Gödel to achieve self-reference. However, there are simpler methods—subsequently discovered by the logicians Alfred Tarski, Willard Quine, and Raymond Smullyan—one of which I will now show you.

"In a paper called *Languages in Which Self-Reference Is Possible*, Smullyan defines the *norm* of an expression to be the expression *followed by* its own quotation. Let us consider an example. We begin with the following expression:

(9) JOHN IS READING

"Its norm is:

(10) JOHN IS READING "JOHN IS READING"

"Sentence (10) is not self-referential; it doesn't say that John is reading (10), but that John is reading (9). But now, let's consider the following instead of (9).

(11) JOHN IS READING THE NORM OF

"Its norm is:

(12) JOHN IS READING THE NORM OF "JOHN IS READING THE NORM OF"

"Sentence (12) is self-referential; it asserts that John is reading the norm of (11), but the norm of (11) is (12) itself.

"Let's look at the symbolic version. As before, we will use J to abbreviate 'John is reading,' and we will now use N to abbreviate 'the norm of.' So (11) is abbreviated thus:

(13) JN

"And (12) is abbreviated thus:

(14) JN "JN"

"Now, (14) is the norm of (13). Also (14) asserts that John is reading the norm of (13). Thus (14) asserts that John is reading (14), so (14) is a self-referential sentence.

"Compare the self-referential sentence JN 'JN,' which is based on normalization, with the sentence JD 'JDx' which is based on diagonalization and which came earlier.

"Using diagonalization, can we construct an expression that denotes itself? Yes; such an expression is D'Dx' which denotes the diagonalization of Dx and the diagonalization of Dx is D'Dx'. Thus D'Dx' denotes itself.

"But using normalization, the solution is even simpler: The expression N'N' denotes the norm of the letter N, which is N'N.' Thus N'N' denotes itself. Of course this is the same as NQ_1NQ_2 of Professor Quincy's system, only using Q_1 in place of the opening quote and Q_2 in place of the closing quote. That's what Quincy meant when he said that his system was based on two-sided quotation."

"Oh, I'm really glad to understand that now," said Annabelle. "I was actually puzzled when he said that, but I didn't wish to interrupt him. He also mentioned systems based on one-sided quotation. What are they?"

"One uses opening quotes but not closing quotes," said the Sorcerer. "It has certain advantages and certain disadvantages over two-sided quotation. One takes a symbol—not the usual one for opening quotes—say the symbol °, and given an expression x, one uses °x instead of x as a name for x. As I said, there are certain advantages to this, which I will shortly explain. Certain machine languages such as LISP use one-sided quotation.

"Following Professor McCulloch's terminology (from Smullyan's book *The Lady or the Tiger?*), we will define the *associ-*

ate of an expression to be the expression, followed by an aster-isk (the symbol "*"), followed again by the expression. And now, let us use the symbol A to abbreviate 'the associate of.' Then, instead of JN'JN,' as a sentence that asserts that John is reading it, we can use the following:

(15) JA*JA

"Sentence (15) asserts that John is reading the associate of JA, but the associate of JA is the very sentence JA*JA. And so we have another method of achieving self-reference.

"Also, an expression that denotes itself is A*A. It denotes the associate of A, which is A*A.

"An advantage of one-sided quotation is that it affords a relatively easy method of achieving *cross-reference*—the con-struction of two sentences each of which talks about the other. Suppose, for example, we consider two individuals—John and Paul—and we wish to construct two sentences x and y such that x says that John is reading y, and y says that Paul is reading x. How can this be done? Let's continue to let J abbreviate 'John is reading,' and let's use P to abbreviate 'Paul is reading.' Now, how can the sentences x and y be constructed?"

· 1 ·

How can they? There are two solutions.

· 2 ·

Suppose now we consider also a third person—call him Wil-liam—and use the letter W to abbreviate "William is reading."

Construct sentences x, y, and z such that x says that John is reading y, y says that Paul is reading z, and z says that William is reading x.

"This notion of *associate,*" said the Sorcerer, "was the basis of several of McCulloch's systems reported in *The Lady or the Tiger?* Now, it so happens that the operation of *repeating* can be used for achieving self-reference and cross-reference just as well as the operation of taking the associate. I don't believe that Smullyan was aware of this when he wrote *The Lady or the Tiger?*, but he demonstrated this in a later paper called *Quotation and Self-Reference.* Now let's take a look at the ideas involved.

"By the *repeat* of an expression we mean the expression followed by itself. Let us use the letter R to abbreviate 'the repeat of.' Using the symbols J, R, and °, how do we construct a sentence that asserts that John is reading it? A common wrong guess is JR°JR. This is wrong because JR°JR says that John is reading the repeat of JR, which is JRJR. Thus JR°JR says that John is reading JRJR; it doesn't say that John is reading JR°JR.

"The correct solution is JR°JR°. This sentence says that John is reading the repeat of JR°, which is the very sentence JR°JR°.

"Let us now compare the four ways we now have of constructing a sentence that asserts that John is reading it. We can use either diagonalization, normalization, association, or repetition, and we respectively have the following four sentences.

(i) JD"JDx"
(ii) JN"JN"
(iii) JA°JA
(iv) JR°JR°

"Of course the repetition method is the whole basis of Charles Roberts's programming system [Chapter 11]. I don't know whether Roberts thought of this himself or got the idea from Smullyan.

"Also, repetition can be used as well as association to achieve cross-reference."

• 3 •

Using the symbols J, P, R, °, construct sentences x and y such that x asserts that John is reading y and y asserts that Paul is reading x.

SELF-REFERENCE USING GÖDEL NUMBERING

"I have heard the phrase *Gödel numbering*," said Annabelle, "and have been told that Gödel used it to achieve self-reference. Could you explain this to us?"

"Certainly. You see, in the mathematical systems studied by Gödel, the sentences talk about things like numbers and sets, not about sentences. The systems have nothing like quotation marks or other devices for talking directly about expressions. And so Gödel cleverly got around this by assigning to each sentence a number, called the *Gödel number* of the sentence. Then, roughly speaking, the Gödel number of the sentence played the role of the quotation of the sentence.

"As a crude illustration, suppose I assign numbers to all sentences in the English language and can find a number n that is the Gödel number of the following sentence.

John is reading the sentence whose Gödel number is n.

"But n is the Gödel number of the very sentence, and so this sentence is saying in a roundabout way that John is reading it.

"Now, how can this be done? I will show you two ways—

the first uses Gödel's diagonalization method. Let's go back to using J for 'John is reading,' and D for 'the diagonalization of,' but a new meaning will now be used for the word 'diagonalization.'

"Let's use the five symbols J, D, x, 1, 0. And let's assign to these five symbols the respective Gödel numbers 10, 100, 1000, 10000, 100000. For easier reference, let me rewrite these five symbols with their Gödel numbers underneath them.

J	D	x	1	0
10	100	1000	10000	100000

"Then, to any compound expression made up of these five symbols, if we replace each symbol by its Gödel number, the resulting number will be the Gödel number of the expression. For example, the expression xJ1D has Gödel number 10001010000100; the expression DJ has Gödel number 10010.

"We now redefine the *diagonalization* of an expression to be the result of replacing the symbol x not by the quotation of the expression (there isn't any in this system), but by the Gödel number of the expression (written, of course, in ordinary base 10 notation). For example, the diagonalization of Jx is J101000. The diagonalization of DxJ is D100100010J.

"And now, for any Gödel number n, we interpret Jn to mean not that John is reading the number n, but that John is reading the *expression* whose Gödel number is n. For example, J10000100 asserts that John is reading the expression '1D.' Or, J10 asserts that John is reading the letter J. Or again, J10010 asserts that John is reading DJ.

"And now we interpret JDn to mean that John is reading the *diagonalization* of the expression whose Gödel number is n. For example, JD10100010 asserts that John is reading the diagonalization of the expression whose Gödel number is 10100010.

Well, the expression whose Gödel number is 10100010 is JxJ, and the diagonalization of JxJ is J10100010J. And so, JD10100010 asserts that John is reading the (meaningless) expression J10100010J.

"And now it should be obvious how to construct a sentence that asserts that John is reading that very sentence."

• 4 •

Find the sentence.

"Now, the principle of Gödel numbering can be used with normalization instead of diagonalization, as was done by Smullyan. He did something like the following.

"Let us now use the four symbols J N 1 0—we no longer need x. Let's assign them the respective Gödel numbers 10, 100, 1000, 10000. Again, the Gödel number of a compound expression is obtained by replacing each of the four symbols by its Gödel number. And now, we redefine the *norm* of an expression to be the expression followed by its Gödel number. For example, the norm of J1JN is J1JN10100010100. We now interpret JNn to mean that John is reading the *norm* of the expression whose Gödel number is n. (If n is not the Gödel number of any expression, we take JNn to be false.)"

• 5 •

Now what simple sentence says that John is reading it?

THE SORCERER'S SPECIAL SYSTEM

"Recently," said the Sorcerer with a rather proud smile, "I thought of another scheme of self-reference which uses neither quotation marks nor Gödel numbering, and that works very neatly for cross-reference as well as self-reference. It is remarkably similar to Simon Simpson's method of programming.

"For any expression x, I shall now write Jx to mean that John is reading the very expression x. I won't put quotes around x, nor a star before the x, nor use the Gödel number of x. I shall boldly write Jx to mean that John is reading x. And now I shall write J̇x to mean that John is reading the *repeat* of x. Thus J̇x means that John is reading xx. (Also, Jxx means the same thing.)

"Self-reference is now completely trivial; the sentence J̇J̇ asserts that John is reading the repeat of J̇, which is J̇J̇. Thus J̇J̇ says that John is reading the very sentence J̇J̇. That's about as simple a method of getting self-reference as can be imagined.

"Cross-reference, though not quite that simple, is still relatively so compared to the other methods we have discussed. I will write Px to mean that Paul is reading x, and Ṗx to mean that Paul is reading xx. And likewise, for William, I will use W and Ẇ similarly. Then, can you see how to construct sentences x and y such that x says that John is reading y, and y says that Paul is reading x? And can you see how to construct sentences x, y, and z such that x says that John is reading y, y says that Paul is reading z, and z says that William is reading x?"

· 6 ·

What are the solutions?

Solutions

1. *One solution:*   x = J°PA°J°PA   y = PA°J°PA
   Another solution: x = JA°P°JA   y = P°JA°P°JA

2. *One solution:*

$$x = J°P°WA°J°P°WA$$
$$y = \qquad\qquad P°WA°J°P°WA$$
$$z = \qquad\qquad\quad WA°J°P°WA$$

(There are at least two other solutions.)

3. *One solution:*   x = J°PR°J°PR°   y = PR°J°PR°
   Another solution: x = JR°P°JR°   y = P°JR°P°JR°

4. JD101001000

5. JN10100

6. First problem is to construct x and y such that x asserts that John is reading y and y asserts that Paul is reading x.

   One solution: x = JṖJṖ   y = ṖJṖ
   Second solution: x = J̇PJ̇   y = PJ̇PJ̇

   This is certainly very economical—clean and simple—compared to the other methods we have seen. My hat is off to the Sorcerer!

   Now for the Sorcerer's 3-cycle problem. There are the following three solutions:

x: JPẆJPẆ   J̇PWJṖ   J̇PWJ̇
y: PẆJPẆ   ṖWJṖ   PWJ̇PWJ̇
z: ẆJPẆ   WJṖWJṖ   WJ̇PWJ̇

15

THE SORCERER'S MINIATURE GÖDELIAN LANGUAGE

"TODAY," said the Sorcerer, "I want to show you a miniature version of Gödel's famous incompleteness theorem. It will serve as a bridge from what we did last time to what we'll get into a bit later. The system I will now present is a modernized and streamlined version of a Smullyan 'language.' I shall use the quotationless method I showed you last time in place of Smullyan's one-sided quotation.

"In the system, various sentences can be proved. The system uses the four symbols P, Ṗ, Q, Q̇. The symbol P means provability in the system—thus, for any expression X in the language of the system, PX asserts that X is provable in the system and accordingly will be called *true* if and only if X is provable in the system. The symbol Q stands for nonprovability in the system and for any expression X, QX asserts that X is not provable in the system and QX is called *true* just in the case that X is not provable in the system. Next, ṖX means that XX is provable in the system, and is accordingly true if and only if this is the case. Lastly, Q̇X means that XX is not provable in the system, and is called *true* if and only if XX is not provable in the system. By a *sentence* is meant any expression of one of the four forms PX, ṖX, QX, Q̇X, where X is any combination of the four symbols. I henceforth use the word

167

provable to mean provable in the system. Let us review the basic facts.

(1) PX asserts that X is provable.

(2) QX asserts that X is not provable.

(3) ṖX asserts that XX is provable.

(4) Q̇X asserts that XX is not provable.

"We see that the system is self-referential in that it proves various sentences that assert what the system can and cannot prove. We are given that the system is wholly accurate in that every sentence provable in the system is true—in other words the following four conditions hold (where X is any expression).

C_1: If PX is provable so is X.

C_2: If QX is provable then X is not provable.

C_3: If ṖX is provable so is XX.

C_4: If Q̇X is provable then XX is not provable.

"Now, just because every sentence provable in the system is true, it doesn't necessarily follow that every true sentence is provable in the system. As a matter of fact, there happens to be a sentence that is true but not provable in the system. Can you find it?"

· 1 ·

Find a true sentence that is not provable in the system.

Refutable Sentences "For each sentence, we define its *conjugate* as follows. The conjugate of PX is QX, and the conjugate of QX is PX. The conjugate of ṖX is Q̇X and the conjugate of Q̇X is ṖX. Thus the sentences PX and QX are conjugates of each other, and the sentences ṖX and Q̇X are conjugates of each other. Given any conjugate pair, it is obvious that one of the pair is true and the other false.

"A sentence is called *refutable* (in the system) if its conjugate is provable (in the system). Thus, PX is refutable if and only if

QX is provable, and PX is provable if and only if QX is refutable. Likewise with ṖX and Q̇X."

· 2 ·

Find a sentence that asserts that it is refutable.

· 3 ·

Find a sentence that asserts that it is not refutable.

· 4 ·

What sentence asserts that it is provable?

Undecidable Sentences. "A sentence is called *undecidable* (in the system) if it is neither provable nor refutable in the system," said the Sorcerer. "Now, as you saw in the solution of Problem 1, the sentence Q̇Q̇ is true but not provable in the system. Since it is true, then its conjugate ṖQ̇ is false, hence also not provable in the system. Thus the sentence Q̇Q̇ is undecidable in the system.

"My argument has appealed to the notion of truth, but even without appeal to this notion one can obtain the undecidability of Q̇Q̇ as a direct consequence of conditions C_1 through C_4 as follows: Suppose Q̇Q̇ were provable. Then by C_4, taking Q̇ for X, the repeat of Q̇ is not provable, which means that Q̇Q̇ is not provable. So if Q̇Q̇ is provable, then it is not provable, which is a contradiction. Therefore, Q̇Q̇ is not provable. If its conjugate ṖQ̇ were provable, then by C_3 (taking Q̇ for X) Q̇Q̇ would be provable, which we just saw is not the case. And so ṖQ̇ is not provable either. Thus the sentence Q̇Q̇ is undecidable in the system."

"Tell me this," said Annabelle. "Is $\dot{Q}\dot{Q}$ the only sentence that is true but unprovable, or are there others?"

"The sentence $\dot{Q}\dot{Q}$," replied the Sorcerer, "is the only sentence that I know of having the property that for *every* system satisfying conditions C_1 through C_4, it is true for that system and unprovable in that system. But, as you will see later, for any system satisfying C_1 through C_4, there are other sentences that are true but unprovable in that system. The sentence $\dot{Q}\dot{Q}$ is, as I said, the only sentence that I know that simultaneously works for *all* systems satisfying C_1 through C_4."

The Sorcerer then gave the following problems:

5 • Some Fixed-Point Properties

Show that for any expression E there is a sentence X that asserts that EX is provable (X is true if and only if EX is provable) and there is some X that asserts that EX is not provable.

6 • Some Anti–Fixed-Point Properties

For any sentence X, let $\overline{X}$ be the conjugate of X.

Show that for any expression E there is some sentence X that asserts that $E\overline{X}$ is provable and a sentence X that asserts that $E\overline{X}$ is not provable.

Next, the Sorcerer presented some problems in cross-reference.

• 7 •

Find sentences X and Y such that X asserts that Y is provable and Y asserts that X is not provable. (There are two solutions.)

Then show that at least one of the sentences X, Y must be true but not provable (though there is no way to tell which).

• 8 •

Now find sentences X and Y such that X asserts that Y is refutable and Y asserts that X is not refutable. (There are two solutions.) Then show that at least one of the two must be false but not refutable (though there is no way to tell which).

• 9 •

Find sentences X and Y such that X asserts that Y is provable and Y asserts that X is refutable. (There are two solutions.) Then show that one of them is true and not provable, or the other is false but not refutable. Which of X, Y is which?

• 10 •

Find sentences X and Y such that X asserts that Y is not provable and Y asserts that X is not refutable. Does it follow that one of them must be undecidable?

• 11 •

Find sentences X, Y, and Z such that X asserts that Y is refutable, Y asserts that Z is not refutable and Z asserts that X is provable. Is one of the three necessarily undecidable?

• 12 •

"I have said before," said the Sorcerer, "that for any system satisfying conditions C_1 through C_4, there are sentences other than $\dot{Q}\dot{Q}$ that are true but unprovable in the system. You are now in a position to prove this. Do you see how?"

Regularity. "I shall call a system *regular*," said the Sorcerer, "if the converses of conditions C_1 and C_3 hold—that is, if X is provable, so is PX and if XX is provable, so is $\dot{P}X$. This together with C_1 and C_3 tell us that PX is provable if and only if X is provable, and $\dot{P}X$ is provable if and only if XX is provable. I might remark that regularity is the analogue of a condition that does hold for the type of systems studied by Gödel, but I'll say more about that another time. Regular systems have some interesting properties, as you will soon see.

"Let me define a *positive* sentence as one of the form PX or $\dot{P}X$ and a *negative* sentence as one of the form QX or $\dot{Q}X$. Positive sentences assert that certain sentences are provable; negative sentences assert that certain sentences are not provable. Let us now note that if the system is regular, then all true positive sentences are provable, and conversely, if all true positive sentences are provable, then the system is regular."

· 13 ·

Why is it that the system is regular if and only if all true positive sentences are provable?

"And so," continued the Sorcerer, "we see that in a regular system, only negative sentences can be true but unprovable. Any sentence that asserts that something *is* provable, if true, must itself be provable."

· 14 ·

If a system is regular, does it necessarily follow that every false negative sentence is refutable?

"Regular systems have some interesting features," said the Sorcerer, "as you will now see."

• 15 •

"For one thing, in a regular system, the ambiguities of Problems 7 through 10 disappear—that is, if we assume the system is regular, then in Problem 7 we can tell whether it is X or Y that is true but unprovable. Which is it? And in Problem 8, is it X or Y that is false but not refutable? And for Problem 9, is it X that is true but provable, or is it Y that is false but not refutable? And for Problem 10, is it X that is undecidable? All this, of course, with the assumption of regularity."

"Let us note," said the Sorcerer, "that for any system satisfying our given conditions C_1 through C_4, whether the system is regular or not, if E is any string of P's, then if EX is provable, so is X. This follows by repeated applications of C_1. For example, if PPPX is provable, so is PPX (by C_1); hence so is PX (again by C_1); hence so is X (again by C_1). You can readily see that the same holds if E contains four or more P's—or if E contains two P's or just one P. And so if E is *any* string of P's, if EX is provable, so is X. For a regular system, the converse also holds—that is, if X is provable, so is EX, where E is any string of P's. For if X is provable and the system is regular, then PX is provable (by regularity); hence so is PPX, and so forth. And so for a regular system, if E is any string of P's, then EX is provable if and only if X is provable.

"Another thing about regular systems is this: For any system satisfying C_1 through C_4, PX is true if and only if PXX is true, for each is true if and only if XX is provable. However, without regularity, there is no reason to believe that PX is provable if and only if PXX is provable. If either one is provable, then the other one is true, but that doesn't mean that the other one is provable. However, if the system is regular, then $\dot{P}X$ is provable if and only if PXX is provable."

· 16 ·

Why is it that in a regular system, $\dot{P}X$ is provable if and only if PXX is provable?

"Now comes a particularly interesting thing about regular systems," said the Sorcerer. "We have already seen that in any system satisfying conditions C_1 through C_4, there are infinitely many sentences that are true for the system but not provable in the system. But this does not mean that there are infinitely many sentences such that each one is true for *all* systems satisfying C_1 through C_4 and at the same time unprovable for *all* such systems. However, there are infinitely many sentences X such that for every *regular* system satisfying C_1 through C_4, each X is true for the system but unprovable in the system."

· 17 ·

Can you prove this?

"What I have shown you today," said the Sorcerer, "has applications to the field known as *metamathematics*—the theory of mathematical systems. My miniature system provides one approach to Gödel's famous incompleteness theorem:
"Let us consider a mathematical system (M) in which there are well-defined rules specifying certain sentences as true and others as provable in (M), and suppose that we wish to know whether (M) is *complete* in the sense that all true sentences of (M) are provable in (M). Now it can be shown that if (M) is any one of a wide variety of systems investigated by Kurt Gödel, it

is possible to *translate* my system into (M) in the sense that corresponding to each sentence X of my system, there is a sentence X* of the system (M) such that X is true in my system if and only if the corresponding sentence X* of (M) is a true sentence of (M), and also, X is provable in my system if and only if X* is provable in (M). Do you realize the ramifications of this? It means that for every such system (M), there must be a true sentence of (M) that is not provable in (M)—its truth can be known only by going outside of the system. Thus, no system (M) into which my system is translatable can possibly be complete. Do you see why this is so?"

• 18 •

Why is this so?

All this is most remarkable!" said Annabelle.

"It certainly is!" agreed Alexander.

"What will you tell us about next time?" asked Annabelle.

"On your next visit," replied the Sorcerer with a mischievous smile, "I have a very special paradox prepared for you."

"I'm looking forward to it," said Annabelle. "I've always been intrigued by paradoxes."

Solutions

1. The sentence is $\dot{Q}\dot{Q}$. It asserts that the repeat of $\dot{Q}$ is not provable, but the repeat of $\dot{Q}$ is $\dot{Q}\dot{Q}$. Hence $\dot{Q}\dot{Q}$ is true if and only if it is not provable in the system. This means that it is either true and not provable or false and provable. The latter alternative contradicts the given condition that only true sentences are provable in the system. Therefore, the former alter-

native holds—the sentence is true but not provable in the system. (This sentence is a transcription of Gödel's famous sentence that asserts its own nonprovability.)

2. $\dot{P}\dot{Q}$ asserts that the repeat of $\dot{Q}$—which is $\dot{Q}\dot{Q}$—is provable. But $\dot{Q}\dot{Q}$ is the conjugate of $\dot{P}\dot{Q}$. And so $\dot{P}\dot{Q}$ asserts that its conjugate is provable, or, what is the same thing, that it is itself refutable.

3. The sentence is $\dot{Q}\dot{P}$. It asserts that the sentence $\dot{P}\dot{P}$—which is the conjugate of $\dot{Q}\dot{P}$—is provable.

4. $\dot{P}\dot{P}$ asserts that it is provable.

5. A sentence X that asserts that EX is provable is $\dot{P}E\dot{P}$, which asserts that the repeat of $E\dot{P}$ is provable. But the repeat of $E\dot{P}$ is $E\dot{P}E\dot{P}$, which is EX.

A sentence X that asserts that EX is not provable is $\dot{Q}E\dot{Q}$.

6. A sentence X that asserts that $E\overline{X}$ is provable is $\dot{P}E\dot{Q}$.

A sentence X that asserts that $E\overline{X}$ is not provable is $\dot{Q}E\dot{P}$.

7. One solution can be obtained by taking X such that it asserts that QX is provable and then taking Y = QX (which asserts that X is not provable). This gives the solution

$$X = \dot{P}Q\dot{P}, \quad Y = \dot{Q}\dot{P}Q\dot{P}$$

Another solution can be obtained by taking some Y that asserts that PY is not provable—namely, Y = $\dot{Q}P\dot{Q}$ and then taking X = PY. We thus have the alternative solution

$$X = P\dot{Q}P\dot{Q}, \quad Y = \dot{Q}P\dot{Q}$$

In either solution X asserts that Y is provable and Y asserts that X is not provable. Therefore, X is true if and only if Y is provable and Y is true if and only if X is not provable. Now, if X and Y are *any* two sentences bearing these two relationships with each other, one of them must be true but unprovable by the following argument: Suppose X is provable. Then X is true; hence Y is provable, hence true; hence X is not provable, which is a contradiction. Therefore, X cannot be provable. It then follows that Y must be true. And so, X is definitely not provable and Y is definitely true. Now, either X is true or it isn't. If X is true, then X is true but not provable. If X is not true, then Y is not provable (because X is true if and only if Y *is* provable); hence Y is then true but not provable.

In summary, X is not provable and Y is true. If X is true, then it is X that is true but not provable; if X is false, then it is Y that is true but not provable.

8. Solution 1: $X = P\dot{P}Q\dot{P}$ $Y = \dot{Q}Q\dot{P}$
    Solution 2: $X = \dot{P}P\dot{Q}$ $Y = Q\dot{Q}PQ\dot{}$

(I obtained these solutions by taking for X the conjugate of the Y of the last problem, and for Y, the conjugate of the X of the last problem.)

Now, X asserts that Y is refutable; hence X asserts that $\overline{Y}$ is provable; hence $\overline{X}$ asserts that $\overline{Y}$ is not provable. Also, Y asserts that X is not refutable; hence Y asserts that $\overline{X}$ is not provable; hence $\overline{Y}$ asserts that $\overline{X}$ is provable. Then by the last problem, taking $\overline{Y}$ for X and $\overline{X}$ for Y, we see that at least one of $\overline{X}$, $\overline{Y}$ is true but not provable; hence one of X, Y is false but not refutable. (Of course, we could have proved this from scratch—and if the reader has any doubts, he or she might try it—but why duplicate work already done?)

9. One way is to take an X that asserts that $P\overline{X}$ is provable and then take $Y = P\overline{X}$. Another is to take some Y that asserts that QY is provable and then take $X = PY$. We thus get the following solutions:

Solution 1: $X = \dot{P}P\dot{Q}$ $Y = P\dot{Q}P\dot{Q}$
Solution 2: $X = P\dot{P}Q\dot{P}$ $Y = \dot{P}Q\dot{P}$

Suppose X is true. Then Y is provable (as X asserts); hence Y is true; hence X is refutable (as Y asserts), which is not possible. Therefore, X cannot be true; it must be false. Then Y is not provable (as X asserts), and so we now know that X is false and Y is not provable. If X is not refutable, then X is false but not refutable. On the other hand, if X is refutable, then what Y asserts is true; hence Y is then true but not provable. And so, either X is false but not refutable, or Y is true but not provable.

10. We can simply take the conjugates of the X and Y of the last problem and interchange them, thus getting the solutions:

Solution 1: X = QQ̇PQ̇   Y = Q̇PQ̇
Solution 2: X = Q̇QṖ   Y = QṖQṖ

By applying our analysis of the last problem to $\overline{Y}$ and $\overline{X}$, instead of X and Y respectively, we see that either $\overline{Y}$ is false but not refutable, or $\overline{X}$ is true but not provable. This means that either X is false but not refutable or Y is true but not provable.

11. We shall give but one solution, which we get by taking an X that asserts that PQX is provable, then taking Y to be QQX and Z to be PX. Thus X asserts that Y is refutable, Y asserts that QX is not provable, or equivalently that PX is not refutable, but PX is Z. Of course, Z asserts that X is provable. We thus have the following solution:

X = ṖPQṖ   Y = QQṖPQṖ   Z = PṖPQṖ

Now, suppose Z is true. Then X is provable, hence true; hence Y is refutable, hence false; hence Z is refutable, hence false; and we have a contradiction. Therefore, Z cannot be true; Z is false. Also then, X is not provable. If X is true, then X is true but not provable. Suppose X is false. Then Y is not refutable. If Y is false, then Y is false but not refutable. If Y is true, then Z is not refutable. And so, one of the following three things must hold: (1) X is true but not provable; (2) Y is false but not refutable; (3) Z is false but not refutable.

12. This is immediate from Problem 7. We know that in any system satisfying conditions C_1 through C_4, at least one of the two sentences ṖQṖ, QṖQṖ is true but not provable in the system. Likewise, with the two sentences PQ̇PQ̇, Q̇PQ̇. And, of course, our old standby Q̇Q̇ is true but not provable in the system. And so, the system contains at least three sentences that are true but not provable in the system. (Actually, there are infinitely many—for example, one of the three sentences PPQ̇PPQ̇, PQ̇PPQ̇, Q̇PPQ̇ must be true but not provable. Also, one of the four sentences PPPQ̇PPPQ̇, PPQ̇PPPQ̇, PQ̇PPPQ̇, Q̇PPPQ̇ is true but unprovable, and so forth.)

13. Suppose the system is regular. Consider a positive sentence X. It is either of the form PY or ṖY. If PY is true, then Y is provable; hence by regularity PY is provable. If ṖY is true, then YY is provable; hence by regularity, ṖY is provable. This shows that regularity implies that all true positive sentences are provable.

Conversely, suppose that all true positive sentences are provable. Well, suppose Y is provable. Then PY is true, hence provable (by hypothesis). Also, if YY is provable, ṖY is true, hence provable (by hypothesis), and so the system is regular.

14. Of course it does! Suppose the system is regular. Now let X be any false negative sentence. Then its conjugate $\overline{X}$ is a true positive sentence, hence provable. Hence, X is refutable.

15. Suppose the system is regular. Then, as we have shown, all true positive sentences are provable in the system and all false negative sentences are refutable.

In Problem 7, X is a positive sentence; hence it is not possible that X is true and unprovable; hence it is definitely Y that is true and unprovable. This goes for either of the two solutions for X and Y. Thus, the sentences QṖQṖ and Q̇PQ̇ are both true and unprovable in every *regular* system.

In Problem 8, it cannot be Y that is false but not refutable, because Y is a negative sentence, so it must be X.

In Problem 9, it cannot be that Y is true and not provable, since Y is a positive sentence, so it must be that X is false but not refutable.

In Problem 10, it cannot be X that is false and not refutable, since X is a negative sentence, so it is Y that is true but not provable.

16. Suppose the system is regular. Then ṖX is provable if and only if ṖX is true, which in turn is the case if and only if XX is provable, which in turn is the case if and only if PXX is true, which in turn is the case if and only if PXX is provable.

More simply, we know that ṖX is true if and only if PXX is true (the truth of either is equivalent to the provability of XX), but ṖX and PXX are both positive sentences, and for a regular system, truth and provability coincide for positive sentences.

17. If E is any string of P's, the sentence Q̇EQ̇ must be both true and unprovable for *all* regular systems. To begin with, even for a system not necessarily regular, EQ̇EQ̇ cannot be provable, for if it were, Q̇EQ̇ would be provable (by repeated applications of C_1); hence EQ̇EQ̇ would not be provable (by C_3), and we would have a contradiction. Therefore, EQ̇EQ̇ is not provable. Hence also Q̇EQ̇ must be true. But now, if we add the assumption that the system is regular, then if Q̇EQ̇ were provable, then EQ̇EQ̇ would also be provable, which is not the case. Therefore, Q̇EQ̇ is not provable in any regular system—yet it is true for all regular systems (satisfying C_1 through C_4, of course).

18.   Since there is a true sentence X of the Sorcerer's system that is not provable in his system (for example, the sentence Q̇Q̇, as we saw in the solution of Problem 1), then its translation X into (M) must be a true sentence of (M) that is not provable in (M).

PART V

HOW CAN THESE THINGS BE?

16

SOMETHING TO THINK ABOUT!

ON THEIR NEXT VISIT, Annabelle and Alexander were introduced to the most baffling paradox they had ever heard in their lives!

"Before I give you this paradox," said the Sorcerer, "I would first like to offer you a little problem in probability that causes much controversy. Many people give the wrong answer and insist they are right, and no argument can convince them they are wrong. Would this interest you?"

"Of course," they both replied.

"Well, the problem is this: You have three closed boxes—call them A, B, and C. One of them contains a prize and the other two are empty. You pick one of the three boxes at random— say, Box A. Before you open it to see whether you have won the prize, I open one of the other two boxes—say, Box B—and show you that it is empty. Then you are given the option of keeping the contents of box A, the box you are holding, or of trading it for the contents of the third box—Box C. The question is: Is there any probabilistic advantage in your trading Box A for Box C or not?"

After some thought, Annabelle replied: "No, it makes no difference whether I trade it or not. When I first pick Box A, the chances are one out of three that my box has the prize. But when I see that Box B is empty, the chances are then one out of two that my box has the prize. That is, the chances are now even that the prize is in Box A or in Box C, and so there is neither any advantage nor any disadvantage in my trading."

"I thoroughly agree," said Alexander.

The Sorcerer smiled. "Yes, that's how most people see it," he said, "but they are wrong. It definitely *is* to your advantage to trade. You will be increasing your chances of winning from one out of three to two out of three."

"I can't see that at all!" said Annabelle. "How can that be? Don't you agree that initially the prize is with equal probability in each of the three boxes?"

"Of course," replied the Sorcerer.

"Then once you know it's not in Box B, it is with equal probability in Box A or Box C. Isn't that obvious?"

"No, it's not obvious," replied the Sorcerer. "It's not even true. What you forget is that it was *I* who opened the box. I deliberately opened a box that I knew was empty."

"So what?" said Alexander. "Suppose you had opened Box B *before* we chose Box A and showed us that it was empty. You mean to tell us that the chances are then greater that the prize is in Box C than in Box A?"

"No," replied the Sorcerer. "In that case, the chances are clearly equal."

"You are confusing me more every minute!" cried Alexander. "Look, let me put it this way. Suppose I choose one box—say Box A—and Annabelle chooses another—say Box C. Then you open Box B and show us that it is empty. According to your logic, it is to my advantage to trade boxes with Annabelle, but also it is equally to her advantage to trade boxes with me, and that is clearly absurd!"

"It certainly would be, if it followed from what I said before, but it doesn't. This is a completely different situation! If you choose Box A and your wife chooses Box C and I then open Box B and show you that it is empty, then of course the chances are equal that the prize is in Box A or in Box C."

"I don't see the difference," said Annabelle.

"The difference is that in the second case—when you each choose a box in advance, you leave me no choice as to what remaining box to open—there is only one. One-third of the time I will be *unable* to open Box B and show it is empty, because one-third of the time it won't be. But in the first case, when you both choose Box A and I have the choice of opening for you Box B or Box C and showing it is empty, I can always be sure of showing you an empty box. Regardless of whether your Box A has the prize or not, at least one of the two remaining boxes is empty, and since I know where the prize is, I simply open one of them that I know to be empty. My showing you this empty box doesn't give you the slightest additional information, since I can *always* manage to show you an empty box.

"Now, it would be different," continued the Sorcerer, "if I myself didn't know where the prize was. If I simply opened Box B or Box C at random and you saw that the box was empty, then the probability of Box A having the prize would indeed jump from one-third to one-half, and in that case it would not be to your advantage or disadvantage to trade your box for Box C. Or, to put it another way, suppose you pick Box A. Then a coin is tossed to determine whether Box B or Box C is to be opened—say heads means we open B and tails means we open C. The coin is tossed and heads comes up. Thus Box B is to be opened. But now, mark this well! *Before* Box B is opened, there is a real possibility that the prize will be in it—in fact the chances are one-third that it will be. Then the box is opened and seen to be empty. In that case, you have really gained more information, and the chances are now equal that your Box A contains the prize. But if instead of tossing a coin, *I* have the choice of opening the box, and I know where the prize is, then it is a certainty that the box I choose to open will be empty. If the prize happened to be in Box B, I wouldn't have opened it; I would have opened the empty Box C instead. And so my showing you an empty box gives you no helpful information whatsoever.

"The correct way to look at it is this: You choose Box A, and there is a one-third probability that you are holding the prize. I then deliberately open a box I know to be empty and show you that it is empty. You have gained no new information whatsoever concerning the probability that your box holds the prize; it is still one-third, but you have gained information about Box C; the probability that it has the prize is now two-thirds; before, it was only one-third. And so you definitely should trade."

"I think I begin to see what you are saying," said Annabelle, "but truthfully, I am not yet convinced. I'll have to think about it some more."

"Well, perhaps this will help you," said the Sorcerer. "To illustrate my point more dramatically, "suppose instead of three boxes we have a hundred boxes and just one of them contains a prize. You pick one of the boxes at random. The chances that your box contains the prize are one out of a hundred, isn't it?"

"Certainly," they both agree.

"All right," said the Sorcerer, "that leaves ninety-nine boxes, and I know where the prize is. I then deliberately pick ninety-eight empty boxes, open them, and show you that they are empty. Do you really believe that the chances of your box holding the prize have risen from one out of a hundred to one-half?"

"That's a good way of putting it," said Annabelle, "but I still want to think about it some more."

Discussion. Of course the Sorcerer is right, but it is amazing the number of people who will never be convinced! I am sure that some of you who are reading this will never be convinced. Most of you will, but a few of you won't. Those of you who don't believe the Sorcerer's argument, I would love to play a few dozen games with you using a hundred boxes and giving you odds of ten to one. You would soon find that you are losing your shirt!

THE ENVELOPE PARADOX

"Now I'd like to tell you of a very baffling paradox that has made the rounds these past few years," said the Sorcerer.

"There are two sealed envelopes on the table. You are told that one of them contains twice as much money as the other. You pick one of them and open it to see how much money is inside. Let's say you find $100 in it. Then you are given the option of keeping it or trading it for the other envelope. Now, the other envelope has either twice as much or half as much with equal probability. Thus the chances are equal that the other envelope has $200 or $50, and the chances of your gaining or losing are equal. And so the odds are in your favor if you trade."

"That sounds perfectly logical," said Annabelle.

Alexander agreed.

"But now comes the curious thing," said the Sorcerer. "Before you open the envelope, you know that whatever the amount you find, your reasoning will be the same, and so the rational thing to do is to trade your envelope for the other one immediately, without bothering to open it. For, let n be the amount in the envelope you are holding. Then the other envelope contains either $n/2$ dollars or $2n$ dollars with equal probability, and so you have an equal chance of winning n dollars or losing n/2 dollars, and so it is to your advantage to trade your envelope for the other one. But had you originally chosen the other one, then, by the same reasoning, you should trade it for the one you are not holding, and clearly this is absurd! That is the paradox."

They thought about this for some time.

"I certainly can see the absurdity of the situation," Annabelle finally said, "but I can't figure out the fallacy in the reasoning. What is it?"

"To tell you the truth, I have not yet heard a completely satisfactory answer to your question," said the Sorcerer. "I have given this problem to several experts on probability theory; some were as puzzled as I, and others gave me an explanation in terms of there being no such thing as a probability measure on the infinite set of positive integers. But I suspect that probability is not the heart of the matter, and I have thought of a new version of the paradox which doesn't involve probability at all."

"Oh?" said Annabelle.

"Yes, my version is this: You pick up one of the two envelopes and you decide that you are going to trade it for the other. Either you will gain or you will lose on the trade. I will now prove to you two contradictory propositions:

"*Proposition 1.* The amount that you will gain, if you do gain, is greater than the amount you will lose, if you do lose.

"*Proposition 2.* The amounts are the same.

"Obviously the two propositions cannot both be true, yet I will prove to you that each one is true.

"The proof of the first proposition is essentially one I have already given you. Let n be the number of dollars in the envelope you are now holding. Then the other envelope has either 2n or n/2 dollars."

"With equal probability," said Alexander.

"Probability is now irrelevant," said the Sorcerer. "I want to leave out probability altogether. The important thing now is that the other envelope has either 2n dollars or n/2 dollars; we don't know which."

"All right," said Alexander.

"Then if you gain on the trade, you will gain n dollars, but if you lose on the trade, you will lose n/2 dollars. Since n is greater than n/2, then the amount you gain, if you do gain—which is n—is greater than the amount you will lose, if you do lose—which is n/2. This proves Proposition 1.

"Now for the proof of Proposition 2. Let d be the difference between the amounts in the two envelopes, or what is the same thing, let d be the lesser of the two amounts. If you gain on the trade, you will gain d dollars, and if you lose on the trade, you will lose d dollars. And so the amounts are the same after all. For example, suppose the lesser envelope has $50; then the greater envelope has $100. If you gain on the trade, that means you are holding the lesser and so you will then gain $50; but if you lose on the trade, that means you are holding the $100 envelope and so you will lose $50. Thus $50 is the amount you stand to gain and it is also the amount you stand to lose. The same argument goes for any d that is the lesser of the two amounts. The number d is the amount you stand to gain and is also the amount you stand to lose. This proves Proposition 2; the amounts are equal after all.

"Well," said the Sorcerer to a thoroughly baffled couple, "which of the two propositions is the correct one? They obviously can't both be right!"

Epilogue. It might amuse the reader to know that Annabelle and her husband never did agree on the matter. Annabelle was absolutely sure that Proposition 1 was correct, and Alexander was equally sure that it was Proposition 2.

"But how can you say that?" asked Annabelle. "Suppose you

open your envelope and find $100. Then you know that the other envelope contains either $50 or $200. And so you gain $100 if you do gain, and lose $50, if you do lose. So isn't it obvious that the amount you stand to gain is greater than the amount you stand to lose? Isn't it obvious that $100 is more than $50? How can you possibly have any doubts on the matter?"

"You are looking at it the wrong way," insisted Alexander. "The amounts in the two envelopes are n dollars and 2n dollars for some n whose value we don't know. If you win on the trade, you will be moving up from n to 2n dollars, hence gaining n dollars, and if you lose on the trade, you will be moving down from 2n to n dollars, hence losing n dollars. Clearly the amounts are the same."

The two never did agree on this! Which do *you* believe to be the true one, Proposition 1 or Proposition 2?

17

OF TIME AND CHANGE

THE NEXT TIME our couple climbed the Sorcerer's tower, they found their host in a particularly philosophical mood. He had spent the morning reading early Greek and Chinese philosophy.

"Time is really a strange thing," he said. "Some mystics have claimed that time is unreal, and at times I am inclined to agree with them!"

"I see," said Annabelle. "At *times* you agree with them and at other times you don't, is that it?"

"That's about it," laughed the Sorcerer. "And speaking of time, I must read you a delightful passage I just found in a work by the ancient Chinese philosopher Chuangtse."

There was a beginning. There was a time before that beginning. And there was a time before the time which was before that beginning. There was being. There was non-being. There was a time before that non-being. And there was a time before the time that was before that non-being. Suddenly there is being and there is non-being, but I don't know which of being and non-being is really being or really non-being. I have just said something, but

193

I don't know if what I have said really says something or
says nothing.*

His two listeners had a good laugh—especially at the last
line.

"I'm also reminded of Smullyan's recipe for immortality,"
said the Sorcerer. "Are you familiar with it?"

"No," said Alexander. "This I must hear!"

"It's really very simple: To be immortal, all you have to do
are the following two things: (1) Always tell the truth; never
make any false statements in the future. (2) Just say: 'I will
repeat this sentence tomorrow!' If you do these two things, I
guarantee that you will live forever!"

Of course, the Sorcerer was right: If today you truthfully say:
"I will repeat this sentence tomorrow," then you will repeat
the sentence tomorrow. Assuming you remain truthful tomorrow, then you will repeat the sentence the next day, then the
next day, then the next day, . . .

"Theoretically, it's a perfect plan," said Annabelle, "but it's
hardly the most *practical* plan in the world!"

"It reminds me of the White Knight's plan for getting over
a gate," remarked Alexander.

"What plan is that?" asked the Sorcerer (who evidently had
never read *Alice's Adventures in Wonderland* and *Through the
Looking Glass*).

"As the White Knight explained it," said Alexander, "the
only difficulty is with the feet: the *head* is high enough already.
So you first put your head on the top of the gate—then the
head's high enough; then you stand on your head—then the
feet are high enough, you see; then you're over, you see."

"Very good," laughed the Sorcerer. "And concerning An-

*Wing-Tsit Chan, A Source Book in Chinese Philosophy (Princeton, N.J.: Princeton
University Press, 1963), pp. 185, 186.

nabelle's objection to Smullyan's plan as not being practical, Smullyan presented his plan in a very amusing story of a man in search of immortality who visited a great sage in the East who was said to be an expert on this subject. The sage explained the plan of how to be immortal, but the man objected, as did Annabelle, on the grounds that the plan was not practical. He said to the sage: 'How can I truthfully say that I will repeat this sentence tomorrow when I don't know whether I'll even be alive tomorrow?' 'Oh,' replied the sage, 'you wanted a *practical* solution! No, I'm not very good at practice; I deal only in theory.' "

Annabelle and Alexander had a good laugh over that one.

"Speaking of immortality," said the Sorcerer, "I recall that when I was a lad, my uncle—who took a keen interest in all topics logical and philosophical—gave me a remarkable proof that it is logically impossible for anyone to die. Would you like to hear it?"

"Oh, yes!" they cried simultaneously.

"Well, my uncle put it this way: If a person dies, when does he die? Does he die while he is living or while he is dead? He can't die while he is dead, because once he is dead, he can no longer die—he is already dead. On the other hand, he can't die while he is still alive because he would then be dead and alive simultaneously, which is not possible. Therefore, he can't die at all."

"I would say," said Annabelle, after some thought, "that at the instant he dies, he is neither alive nor dead. The instant is the transition instant between life and death."

"That seems reasonable," said Alexander.

"No," said the Sorcerer, "you can't get out of it that easily! By 'dead' I mean no longer alive. In fact, to avoid any semantic confusion, I'll phrase the argument differently: at the instant of a person's death, is he alive or not alive at that instant?"

They thought some more about this.

"Is this related to Zeno's proofs of the impossibility of motion?" asked Annabelle.

"There is some connection," replied the Sorcerer. "In fact, my uncle's argument can be generalized to give a new proof of the impossibility of motion."

"What are Zeno's arguments?" asked Alexander. "I have heard *of* them, but I have never heard them."

"I have read them," Annabelle said, "but I could never quite figure out where the fallacies lie. There *must* be fallacies, since things obviously *do* move. Just what are the fallacies?"

"Zeno had three proofs," replied the Sorcerer. "There is also a fourth proof, but this proof was poorly recorded and there does not seem to be uniform agreement as to just what that proof is. So I will confine myself to just the first three.

"The first argument is this: Suppose an object moves from a point A to a point B. Before it can get to B, it must get to the point A_1 midway between A and B; call this the *first* step. After completing the first step, the body must get from A_1 to the point A_2 midway between A_1 and B; call this the *second* step. Then it must take the third step—getting from A_2 to the point A_3 midway between A_2 and B, and so forth.

$$\overset{\displaystyle \cdot}{A} \rule{4cm}{0.4pt} \overset{\displaystyle \cdot}{\underset{A_1}{}} \rule{2.5cm}{0.4pt} \overset{\displaystyle \cdot}{\underset{A_2}{}} \rule{1cm}{0.4pt} \overset{\displaystyle \cdot}{\underset{A_3}{}} \overset{\displaystyle \cdot}{\underset{A_4 A_5 B}{}}$$

"After no finite number of steps will the body have reached B; that is, for each positive integer n, after it has performed the first n steps and is at a point A_n, it still has to take another step and go to the point A_{n+1} midway between A_n and B. So the body must take infinitely many steps to get to B and since this is impossible in a finite length of time, the object cannot move from A to B.

"Looking at it another way, before the object can get from A to B, it must first get to the midpoint B_1; but before it can get to B_1, it must first get to the point B_2 midway between A and B_1; but before it can do *that*, it must get to the point B_3 midway between A and B_2, and so forth. Therefore, the body can't even get started!

```
•–•–•—•————•—————————•——————————————————————•
A B₅B₄  B₃      B₂          B₁                              B
```

"Zeno's second argument is about Achilles trying to overtake a tortoise. To begin with, let us say that Achilles is 100 yards behind the tortoise and that he is running ten times as fast as the tortoise. His first step is to run to the spot where the tortoise now is—that is, he runs 100 yards. When he reaches that spot, the tortoise will no longer be there; it will have moved 10 yards forward. Then Achilles takes the second step—he runs 10 yards to reach the spot where the tortoise was at the end of the first step, but when he gets there, the tortoise will have run (or rather walked) 1 yard further. Then when Achilles runs this yard forward, the tortoise will still be $\frac{1}{10}$ of a yard away, and so forth. In other words, whenever Achilles reaches the place where the tortoise had been, the tortoise is no longer there. Hence Achilles can never overtake the tortoise.

"Zeno's third proof strikes me as the most sophisticated; it is the proof about the flying arrow. Suppose an arrow is flying continuously forward during a certain time interval. Take any instant in that time interval. It is impossible that the arrow is moving *during that instant,* because an instant has duration *zero,* and the arrow cannot be at two different places at the same time. Therefore, at every instant the arrow is motionless, hence the arrow is motionless throughout the entire interval, which means that it can't move during the interval!"

"Yes, I remember those arguments, now that you have reminded me of them," said Annabelle, "and I am as much puzzled as ever. Surely, *something* must be wrong with them, since things obviously *do* move, but what is it that is wrong? Is there no logical explanation? Or is there something wrong with logic itself?"

"No, there is certainly nothing wrong with logic itself," laughed the Sorcerer, "and there certainly is a perfectly logical explanation as to just what the fallacies are in Zeno's arguments. I am really surprised that people are fooled by the first two arguments, though it is more understandable with the third, which is more difficult to see through."

"What *are* the fallacies?" asked Alexander.

"Well, let's start with the first one," said the Sorcerer. "In any argument that leads to a false conclusion, there must be a false step somewhere, hence there must be a *first* false step. So, where is the first false step in Zeno's first argument?"

The two of them thought about this for a while.

"I can't see any false step," said Alexander; "every step seems correct to me."

"To me also," said Annabelle.

"Really now!" said the Sorcerer. "Even the conclusion seems correct to you?"

"Of course not!" said Annabelle. "The conclusion is clearly false. Obviously things *do* move."

"Well then, if the conclusion is false and every step before the conclusion is correct, then obviously the first false step is the conclusion itself."

"I never thought of that!" said Annabelle.

"Most people don't, and that's the surprising thing!" said the Sorcerer. Of course, the conclusion is the first wrong statement! Every object must go through the infinite number of steps that Zeno described, but it is completely specious to conclude: 'Therefore the object can't move.' Where is the warrant for that inference? Zeno tacitly assumed that one cannot perform infinitely many steps in a finite length of time, and that assumption is totally unwarranted."

"That assumption seems perfectly reasonable to me!" said

Annabelle. "How can the sum of infinitely many quantities be finite?"

"That happens all the time in mathematics," replied the Sorcerer. "For example, the infinite series $1 + \frac{1}{2} + \frac{1}{4} + \frac{1}{8} + \frac{1}{16} + \frac{1}{32} + \ldots + \frac{1}{2^n} + \ldots$ all adds up to 2. Such an infinite series is called *convergent;* it converges to 2. On the other hand, the infinite series $1 + \frac{1}{2} + \frac{1}{3} + \frac{1}{4} + \frac{1}{5} + \ldots + \frac{1}{n} + \ldots$ is what is called a *divergent* series; if you take enough terms of it, you'll get beyond a million; take enough more terms, you'll get beyond a trillion; whatever number you pick, no matter how large, that series eventually gets beyond that number. Such a divergent series is also said to approach infinity. However, the series relevant to Zeno's first problem is $\frac{1}{2} + \frac{1}{4} + \frac{1}{8} + \ldots$, which converges to 1, and so Zeno was totally unjustified in concluding that the object could never get from A to B. A similar analysis applies to Zeno's second argument—in fact the series is $\frac{1}{10} + \frac{1}{100} + \frac{1}{1000} + \ldots + \frac{1}{10^n} + \ldots$, which converges even more rapidly.

"The third argument of the flying arrow is more subtle, and I am afraid you don't know enough mathematics to understand the resolution. Those who have studied calculus know that at any instant during the interval, the arrow *is* in motion at that instant, and that Zeno was wrong in saying that if the arrow were in motion at an instant, then it would be in more than one place at the instant. Of course, the arrow is only at one place at any given instant, but that does *not* imply that the arrow is at rest at that instant."

"I don't understand," said Annabelle. "If an arrow in motion is only at one place at a given instant, how is it any different from an arrow at rest, which is also at only one place at an instant? How can one tell the difference between the two cases?"

"You ask a great and profound question," replied the Sorcerer, "and the purpose of the differential calculus—or at least one of its main purposes—is to answer just this sort of question. For the first time in the history of mankind, the inventors of the calculus—Newton and Leibniz—gave a perfectly precise definition of what it means for a body to be in motion *at an instant* and what it means to say that the velocity of the body is such and such *at an instant*. I cannot give you these definitions without first teaching you some fundamental notions of the calculus. For now, suffice it to say that the statement that a body has a certain velocity *at an instant* is really a statement not about just the instant but about successively smaller and smaller time intervals centered around the instant. I hope one day I can give you a more complete and technical account of the matter."

"You said before," said Alexander, "that your uncle's argument can be generalized to provide yet another proof of the impossibility of motion. How is that?"

"My uncle's argument can be generalized to show that nothing can ever change—which surely would have pleased the philosopher Parmenides! Suppose something changes from being in one state—call it state A—to being out of state A. At what instant can it make the change? It can't make the change when it is out of state A, because it is already out of state A. And it can't make the change while it is still in state A, because it would then be in state A and out of state A at the same time. And don't tell me that at the instant of change, it is neither in state A nor out of state A, because by *out of state A* I mean *not in state A*, and it is logically impossible for something to be neither in state A nor not in state A at the same time."

"How does that give another proof of the impossibility of motion?" asked Alexander.

"Well, obviously if an object moves away from a given position, it changes its state from being in that position to not being in that position. And so if changing from one state to another is impossible, then motion is impossible. . . .

"Yes, time and change are curious things! What really is time, anyway? As Augustine said: 'When asked what time is, I know not; when asked not, I know!'

"And speaking of time, I understand you will both be leaving us for a while?"

"Yes," replied Annabelle, "tomorrow we sail for home. My sister, Princess Gertrude, is getting married."

"How long will you be gone?" asked the Sorcerer.

"Probably a couple of months," replied Annabelle. "There are several things Alexander and I must attend to back home. But we definitely plan to come back. In fact, we are thinking of moving here. We are both absolutely intrigued by this whole subject of logic. No one at home seems to know much about it. You have no idea how grateful we are for all that you have taught us. But I am beginning to see that we have so much to learn! When we come back, could you tell us something about infinity? That's one subject that has always intrigued and puzzled me!"

"Ah, yes!" said the Sorcerer. "Infinity. That *is* a subject indeed! Yes, when you come back, I promise to give you a guided tour through infinity. Do come back soon, and bon voyage!"

PART VI

A JOURNEY INTO INFINITY

18

WHAT IS INFINITY?

Our couple were away for more than two months. When they finally returned to the island, they lost no time in visiting the Sorcerer.

"Welcome back!" he said. "And so you want to know about infinity?"

"You have a good memory," said Annabelle.

"All right now," said the Sorcerer. "The first thing we should do is to carefully define our terms. Just what is meant by the word 'infinite'?"

"To me, it means endless," said Alexander.

"I would say the same," said Annabelle.

"That's not quite satisfactory," said the Sorcerer. "A circle is endless, in the sense of having no beginning or end, and yet you wouldn't say it's infinite—that is, it has only finite length, though it does have infinitely many points.

"I want to talk about infinity in the precise sense used by mathematicians. Of course, there are other uses of the word; for example, theologians often refer to God as infinite, though some of them are honest enough to admit that when applied to God the word has a different meaning than when applied to anything else. Now, I do not wish to disparage the theological or any other nonmathematical use of the word, but I want to make it clear that what I'm proposing to discuss is infinity in the purely *mathematical* sense of the term. And for this we need a precise definition.

"The word 'infinite' is obviously an adjective, and the first thing we must agree upon is the sort of things to which the adjective is applicable. What sort of things can be classified as either finite or infinite? Well, in the mathematical use of the term, the answer is that it is *sets*, or collections of things, that can be called *finite* or *infinite*. We say that a *set* of objects has finitely many or infinitely many members, and now we must make these notions precise.

"The clue here lies in the notion of a one-to-one correspondence between one set and another. For example, a flock of seven sheep and a grove of seven trees are related to each other in a way that neither is related to a pile of five stones, for the set of seven sheep can be paired with the set of seven trees, for example, by tethering each sheep to a tree, so that each sheep and each tree belongs to exactly one of the pairs. Or, in mathematical terms, a set of seven sheep can be put into a 1 to 1 correspondence with a set of seven trees.

"Another example: Suppose you look into a theater and see that every seat is taken and that no one is standing—and also, that no one is sitting on anyone's lap; there is one and only one person to each seat. Then, without having to count the number of people, or the number of seats, you know that the numbers are the same, because the set of people is in a 1 to 1 correspondence with the set of seats: each person corresponds to the seat he is occupying.

"Now, I know that you are familiar with the set of natural numbers, though you might not be familiar with them by that name. The natural numbers are simply the numbers 0, 1, 2, 3, 4, . . . That is, by a natural number is meant either zero or any positive whole number."

"Is there such a thing as an *unnatural* number?" asked Annabelle.

"No, I have never heard of that," said the Sorcerer, "and I must say the idea strikes me as funny! Anyway, from now on I will be using the word *number* to mean *natural number*, unless I say something to the contrary.

"Now, given a natural number n, just what does it mean to say that a certain set has exactly n members? For example, what does it mean to say that there are exactly five fingers on my right hand? It means that I can put the set of fingers on my right hand into a 1 to 1 correspondence with the set of positive whole numbers from 1 to 5—say, by letting my thumb correspond to 1, the next finger to 2, the middle finger to 3, the next finger to 4, and the little finger to 5. And, in general, given any positive whole number n, we say that a set has (exactly) n members if the set can be put into a 1 to 1 correspondence with the set of positive integers from 1 to n. A set with n elements is also called an $n - element\ set$. And the process of putting an $n -$ element set into a 1 to 1 correspondence with the set of positive integers has a popular name—the popular name of this process is *counting*. Yes, that's exactly what counting is.

"And so, I have told you what it means for a set to have n elements, where n is a *positive* whole number. What about when n is zero; what does it mean for a set to have 0 elements? It obviously means that the set has no elements at all."

"There are such sets?" asked Alexander.

"There is only one such set," replied the Sorcerer. "This set is technically called the *empty set*, and is extremely useful to mathematicians. Without it, exceptions would constantly have to be made and things would get very cumbersome. We want, for example, to be able to speak about the set of people in a theater at a given instant. It may happen that there are no people there at a given instant, in which case we say that the set of people then in the theater is empty—just as we speak of an empty theater. This is not to be confused with no theater at all! The theater still exists as a theater; there just happen to be no people in it. Likewise, the empty set exists as a set, but it has no members.

"I recall a charming incident: Many years ago, I told a lovely lady musician about the empty set. She seemed surprised and said: 'Mathematicians really use this notion?' I replied: 'They certainly do!' She asked: 'Where?' I replied: 'It's used all over the place.' She thought for a moment and said: 'Oh, yes. I guess it's like the rests in music.' I thought that was quite a good analogy!

"Smullyan relates an amusing incident. When he was a graduate student at Princeton University, one of the famous mathematicians there said during a lecture that he hated the empty set. At the next lecture, he made use of the empty set. Smullyan raised his hand and said, 'I thought you said that you disliked the empty set.' The professor replied: 'I said I disliked the empty set. I never said that I don't *use* the empty set!' "

"You haven't yet told us," said Annabelle, "what you mean by 'finite' and 'infinite.' Aren't you going to?"

"I was just about to," replied the Sorcerer. "All the things I've said so far are just leading up to the definition. We say that a set is *finite* if there exists a natural number n such that the set has exactly n elements—which we recall means that the set can be put into a 1 to 1 correspondence with the positive integers from 1 to n. If there is no such natural number n, then the set is called infinite. It's as simple as that. Thus a 0-element set is finite; a 1-element set is finite, a 2-elements set is finite, . . . and an n-element set is finite, where n is any natural number. But if for every natural number n, it is false that the set has exactly n elements, then the set is infinite. Thus, if a set is infinite, then for any natural number n, if we remove n elements from the set, there will be elements left over—in fact, infinitely many elements will be left over.

"Do you see why this is true? Let's first consider a simple problem. Suppose I remove a single element from an infinite set. Is what remains necessarily infinite?"

"It would certainly seem so!" said Annabelle.

"It certainly does!" said Alexander.

"Well, you are right, but can you prove it?"

The two thought about it, but had difficulty proving it. It seemed too obvious to need proof. But it is not difficult to prove this from the very definitions of "finite" and "infinite." These definitions must be used.

· 1 ·

How is this proved?

It took a bit of prodding, but the two finally came up with a proof that satisfied the Sorcerer.

Hilbert's Hotel. "Infinite sets," said the Sorcerer, "have some curious properties that at times have been labeled paradoxical. They are not really paradoxical, but they are a bit startling when first encountered. This is well illustrated by the famous story of Hilbert's Hotel.

"Suppose we have an ordinary hotel with only a finite number of rooms—say a hundred. Suppose all the rooms are taken and each room has one occupant. A new person arrives and wants a room for the night, but neither he nor any of the hundred guests is willing to share a room. Then, it is impossible to accommodate the new arrival; one cannot put 101 people into a 1 to 1 correspondence with 100 rooms. But with infinite hotels (if you can imagine such a thing) the situation is different. Hilbert's Hotel has infinitely many rooms—one for each positive integer. The rooms are numbered consecutively—Room 1, Room 2, Room 3, . . . Room n, . . . and so forth. We can imagine the rooms of the hotel arranged linearly—they start at some definite point and go infinitely far to the right. There is a first room, but there is no last room! It's important to remember that there is no last room—just as there is no last natural number. Now again, we assume that all the rooms are occupied—each room has one guest. A new person arrives on the scene and wants a room. The interesting thing is that it *is* now possible to accommodate him. Neither he nor any of the other guests is willing to share a room, but the guests are all cooperative in that they are willing to change their rooms, if requested to."

• 2 •

How can this be done?

"Now for another problem," said the Sorcerer, after the solution to this last problem had been discussed. "We consider the same hotel as before. But now, instead of just one new person arriving, an infinite number of new guests arrive—one for each positive integer n. Let us call the old guests P_1, P_2, . . . P_n, . . . , and the new arrivals Q_1, Q_2, . . . , Q_n, . . . The new Q-people all want accommodations. The surprising thing is that it can be done!

· 3 ·

How?

"Now for a still more interesting problem," said the Sorcerer. "This time we have infinitely many hotels—one for each positive integer n. The hotels are numbered Hotel 1, Hotel 2, . . . Hotel n, . . . and each has infinitely many rooms—one for each positive integer. The hotels are arranged in a rectangular square array—thus:

Hotel 1	1	2	3	4	5	6	7 . . .
Hotel 2	1	2	3	4	5	6	7 . . .
Hotel 3	1	2	3	4	5	6	7 . . .

· ·

· ·

· ·

· ·

Hotel n	1	2	3	4	5	6	7 . . .

· ·

· ·

· ·

· ·

"The whole chain of hotels is under one management. All the rooms of all the hotels are occupied. One day the management decides to shut down all the hotels but one, in order to save energy. But this means shifting all the inhabitants of all the hotels into just one of the hotels—again, only one person to a room."

· 4 ·

Is this possible?

"You see what these problems reveal," said the Sorcerer. "They show that an infinite set can have the strange property of being able to be put into a 1 to 1 correspondence with a proper part of itself. Let me make this more precise.

"A set A is called a *subset* of a set B if every element of A is also an element of B. For example, if A is the set of numbers from 1 to 100 and B is the set of numbers from 1 to 200, then A is a subset of B. Or if E is the set of all even integers and N is the set of all integers, then E is a subset of N. A subset A of B is called a *proper* subset of B if A is a subset of B but does not contain *all* the elements of B. In other words, A is a proper subset of B if A is a subset of B, but B is not a subset of A. Now, let P be the set $\{1, 2, 3, \ldots n, \ldots\}$ of all positive integers and let P− be the set $\{2, 3, \ldots n, \ldots\}$ of all positive integers other than 1. We have seen in the first problem of Hilbert's Hotel that that P can be put into a 1 to 1 correspondence with P−, yet P− is a *proper* subset of P! Yes, an infinite set can have the strange property of being able to be put into a 1 to 1 correspondence with a *proper* subset of itself! This was known long ago. In 1638, Galileo pointed out that the *squares of the positive integers* can be put in a 1 to 1 correspondence with the positive integers themselves, thus

$$1, \ 4, \ 9, \ 16, \ 25, \quad \ldots n^2, \ldots$$

$$\updownarrow \ \updownarrow \ \updownarrow \ \updownarrow \ \ \updownarrow \qquad \updownarrow$$

$$1, \ 2, \ 3, \ 4, \ \ 5, \quad \ldots, n, \ \ldots$$

"This seemed to contradict the ancient axiom that the whole is greater than any of its parts."

"Well, doesn't it?" asked Alexander.

"Not really," replied the Sorcerer. "Suppose that A is a proper subset of B. Then, in one sense of the word 'greater,' B is greater than A—namely, in the sense that B contains all elements that A contains and some elements that B doesn't contain. But that does not mean that B is *numerically* greater than A."

"I'm not sure I know what you mean by numerically greater," said Annabelle.

"Good point!" said the Sorcerer. "First of all, what do you think it means for one set A to be *of the same size* as another set B?"

"I guess it means that A can be put into a 1 to 1 correspondence with B," said Annabelle.

"Right! And what would you guess it means to say that A is of *smaller* size than B, or that, numerically, A has fewer elements than B?"

"I guess it would mean that A can be put into a 1 to 1 correspondence with a proper subset of B."

"Nice try," said the Sorcerer, "but it won't work. That definition would be fine for finite sets, but not for infinite sets. The trouble is that it might be possible for A to be put into a 1 to 1 correspondence with a proper subset of B, and it might also be possible for B to be put into a 1 to 1 correspondence with a proper subset of A. In which case would you want to say that each is of smaller size than the other? For example, let O be the set of odd positive integers and E be the set of even positive integers. Obviously, O can be put into a 1 to 1 correspondence with E.

$$1, \ 3, \ 5, \ 7, \ 9, \ \ldots 2n - 1 \ldots$$
$$\updownarrow \ \updownarrow \ \updownarrow \ \updownarrow \ \updownarrow \qquad \updownarrow$$
$$2, \ 4, \ 6, \ 8, \ 10, \ \ldots 2n \ldots$$

"But also O can be put into a 1 to 1 correspondence with a *proper* subset of E thus

$$1, \ 3, \ 5, \ 7, \ \ 9, \ \ldots 2n - 1 \ldots$$
$$\updownarrow \ \updownarrow \ \updownarrow \ \updownarrow \ \ \updownarrow \qquad \updownarrow$$
$$4, \ 6, \ 8, \ 10, \ 12, \ \ldots 2n + 2$$

"At the same time, E can be put into a 1 to 1 correspondence with a proper subset of O thus

$$2, \ 4, \ 6, \ 8, \ 10, \ \ldots 2n \ldots$$
$$\updownarrow \ \updownarrow \ \updownarrow \ \updownarrow \ \updownarrow \qquad \updownarrow$$
$$3, \ 5, \ 7, \ 9, \ 11, \ \ldots 2n + 1 \ldots$$

"Now, you certainly wouldn't want to say that E and O are of the same size, yet E is of smaller size than O and also O is of smaller size than E! No, that definition won't serve."

"Then what is the correct definition of 'smaller than' when applied to infinite sets?" asked Annabelle.

"The correct definition is this: We say that A is *smaller in size than B*, or that B is *greater in size than A* if the following two conditions are met: (1) A can be put into a 1 to 1 correspondence with a proper subset of B; (2) A *cannot* be put into a 1 to 1 correspondence with all of B.

"It is crucial that *both* conditions be met," emphasized the Sorcerer, "in order that it can be correctly said that A is smaller than B. To say that A is smaller than B means first of all that A can be put into a 1 to 1 correspondence with a subset of B and also that *every* 1 to 1 correspondence between A and a subset of B must leave out some elements of B.

"And now comes a fundamental question," said the Sorcerer. "Is it the case that any two infinite sets are necessarily of the same size, or do infinite sets come in different sizes? That is the first question to be answered in building up a theory of infinity, and fortunately the question was answered by Georg Cantor towards the end of the last century. The answer created a storm and started a whole new branch of mathematics whose ramifications are fantastic!

"I will give you Cantor's answer at our next session. Meanwhile, do you have a guess as to what the answer is? Are all infinite sets of the same size, or do they come in different sizes?"

Remark. I raise this problem to my students in my beginning logic classes, and among those students who don't know the answer, about half guess that all infinite sets are of the same size and about half guess otherwise.

Those of you who don't already know the answer, would you care to hazard a guess before turning to the next chapter?

Solutions

1. Let us first show that when you add an element to a finite set, you get a finite set. Well, suppose a set A is finite. By definition, this means that for some natural number n, the set A has n elements. If we add a new element to A, the resulting

set then obviously has $n + 1$ elements, hence by definition is finite.

From this, it immediately follows that if we remove an element from an infinite set B, the resulting set must be infinite, for if it were finite, we could put back the element and the resulting set—which is the original set B that we started with—would be finite, which it isn't.

2.   All that the management need do is to request that each inhabitant move one room to the right—in other words, the occupant of Room 1 goes to Room 2, the occupant of Room 2 goes to Room 3, . . . the occupant of Room n moves to Room $n + 1$. Since the hotel has no last room (unlike the more normal finite hotel), no one will be out in the cold. (In a finite hotel, the person in the last room would have no room to his right). After all the guests have been kind enough to move, Room 1 is now empty and the newly arrived guest can take it.

Mathematically, what we have done is to put the set of all positive integers into a 1 to 1 correspondence with the set of all positive integers from 2 on. Of course, the manager of the hotel could have done a similar thing if a hundred million new guests had arrived instead of just one. The manager would then have asked each guest to move a hundred million rooms to the right (the person in Room 1 would move to Room 100,000,001; the person in Room 2 would move to Room 100,000,002, and so forth). For any natural number n, the hotel could accommodate n new guests—just have each guest move n rooms to the right, leaving the first n rooms empty for the new guests.

3.   Now, if infinitely many new guests $Q_1, Q_2, \ldots Q_n, \ldots$ arrive, the solution is a little different. One false solution that has been suggested is that the manager first asks every one of the old guests to move one room to the right. Then, one new guest is put into the empty Room 1. Then the manager again asks everyone to move one room to the right, and so Room 1 is again vacant and a second new guest is put in Room 1. Then this operation is repeated, again and again, infinitely many times, and so sooner or later every new guest gets into the hotel.

But what a horribly restless solution! No one keeps a room permanently and in no finite time would all the guests be settled—infinitely many shiftings would be required. No, the whole thing can be done cleanly with only one shifting. Can you see what this shifting is?

The shifting is that each old guest doubles his room number—that is, the person in Room 1 goes to Room 2, the one in Room 2 goes to Room 4, the one in Room 3 goes to Room 6, ... the one in Room n goes to Room 2n. All these are, of course, done simultaneously, and after the shifting, all the even-numbered rooms are occupied and all the infinitely many odd-numbered rooms are now free. And so, the first new guest Q_1 goes to the first odd-numbered room—Room 1; Q_2 goes to Room 3; Q_3 goes to Room 5, and so forth (Q_n moves to Room $2n - 1$).

4. We first "number" all the occupants of all the rooms of all the hotels according to the following plan:

1		4		9		16	.	.	.
↓		↑		↑		↑	.	.	.
2	→	3		8		15	.	.	.
				↑		↑	.	.	.
5	→	6	→	7		14	.	.	.
						↑	.	.	.
10	→	11	→	12	→	13	.	.	.
.		.		.		.	.	.	.

.
.
.
.

And so, each of the guests is "tagged" with a positive integer. Then all of them vacate their rooms and wait outside for a bit. Then the management shuts down all the hotels but one, and each guest is asked to occupy the room whose number is the same as the number with which he is tagged—the person tagged n goes to Room n.

)

19
CANTOR'S FUNDAMENTAL DISCOVERY

"WELL," SAID THE Sorcerer at the next session, "have you thought about the matter? Have you any guess as to whether there is more than one infinity or only one?"

One of the two (I forget which) guessed one way, and the other, the other.

"The curious thing," said the Sorcerer, "is that Cantor originally conjectured that any two infinite sets must be of the same size, and as I understand it, he spent twelve years trying to prove his conjecture. Then, in the thirteenth year, he found a counterexample—which I like to call a 'Cantor-example.' Yes, there is more than one infinity—there are infinitely many, in fact. And this basic discovery is due to Cantor.

"Now, here is what Cantor did. A set is called *denumerably infinite,* or, more briefly, *denumerable,* if it can be put into a 1 to 1 correspondence with the set of positive integers. And so the question considered by Cantor is: Is every infinite set denumerable? As I said, he first guessed that every infinite set is denumerable, and only later discovered the real truth. What he did, during his initial investigation, was to consider sets that appeared to be too large to be denumerable, but then by some clever device he was able to enumerate them after all."

"What do you mean by *enumerating* a set?" asked Annabelle.

"To *enumerate* a set means to put it into a 1 to 1 correspondence with the set of positive integers. In fact, the word 'enumerable' is used synonymously with 'denumerable.' Anyway, as I said, Cantor considered one set after another which at first sight appeared to be non-denumerable—which means infinite but not denumerable—and found a clever way to enumerate it.

"To illustrate his method, let us imagine that I am the Devil and you are my victims in Hell. That's not too hard to imagine, is it?"

Annabelle and Alexander had a good laugh at that thought.

"Now, I tell you that I will give you a test. I tell you: 'I am thinking of a positive whole number, which I have written down on this folded piece of paper. Each day, you have one and only one guess as to what the number is. If and when you correctly guess it, you can go free, but not before then.' Now, is there some strategy by which you can be sure of getting out sooner or later?"

"Of course," said Alexander. "On the first day I ask if it's 1, on the second day, if it's 2, and so forth. Sooner or later I'm bound to hit your number."

"Right," said the Sorcerer. "Now, my second test is a tiny bit more difficult. This time, I write down either a positive whole number or a negative whole number—I write either one of the numbers 1, 2, 3, 4, . . . or one of the numbers -1, -2, -3, -4, . . . and again each day you have one and only one guess as to what the number is. Now, do you have a strategy that will surely get you out sooner or later?"

"Of course," said Annabelle. "On the first day I ask whether the number is $+1$, on the next day, I ask if it is -1, then I keep going $+2$, -2, $+3$, -3, $+4$, -4, . . . and so on. Sooner or later I must hit your number."

"Right," said the Sorcerer, "and you see what this means.

On the surface, it would seem that the set of positive and negative whole numbers taken together should be larger than—in fact twice the size of—the positive whole numbers alone. Yet, you have just seen how to put the set of positive and negative whole numbers into a 1 to 1 correspondence with the set of positive whole numbers, and so the two sets are really of the same size after all. The set of all positive and negative whole numbers taken together is denumerable. The problem you have just solved is substantially the same as the second problem I gave you concerning Hilbert's Hotel. As you recall, there were denumerably many people in denumerable many rooms, and then a second denumerable set of people came along and we wanted to put those two sets together and accommodate all the members.

"The next test I give my victim is definitely more difficult. This time, I write down *two* numbers on a piece of paper, or maybe the same number written twice. For example, I might write 3, 57, or I might write 17, 206, or I might write 23, 23. Each day, you have one and only one guess as to what the two numbers are. You are not allowed to guess one of them one day and the other another day; you must guess them both on the same day. Now, do you believe there is a strategy by which you can certainly get out sooner or later?"

"I doubt it," said Annabelle. "There are infinitely many possibilities for the first number that you write down, and with each of those possibilities there are infinitely many possibilities for the second number, and so it would seem that an infinity times itself should be bigger than the infinity you start with."

"So it might *seem*," said the Sorcerer, "and so it did seem to many at Cantor's time, but appearance is sometimes deceptive. Yes, there *is* a strategy for certainly getting out. The set of possibilities that you are dealing with is really denumerable after all. Can you find the strategy?"

"Amazing!" said Annabelle, and Alexander agreed. The two

then put their heads together and came up with a simple strategy that definitely works.

• 1 •

What strategy will work?

"Suppose I make the problem slightly harder by now requiring not only that you guess what the two numbers are, but also in what order they are written—say, one is written to the left of the other. Can you now get out for sure?"

"Of course," said Annabelle. "This problem is easy, now that the last problem was solved."

• 2 •

What is the solution?

"Then let me ask you this," said the Sorcerer. "What about the set of all positive fractions? Is that set denumerable or not? You are now in a good position to answer that question. By a *positive fraction* is simply meant a quotient of two positive integers—numbers like $3/7$ or $21/13$."

• 3 •

Is the set of positive fractions denumerable?

"The answer was quite a shock to many mathematicians of Cantor's time," said the Sorcerer. "And now I have a more difficult problem for you. This time, I write down some *finite* set of positive integers. I'm not telling you how many numbers are in the set, nor what the highest number of the set is. Each day, you have one and only one guess as to what the set is. If and when you correctly guess the set, you go free. Now, do you think there is a strategy for going free?"

The two felt that it was quite unlikely.

"There *is* such a strategy," said the Sorcerer. "The set of all *finite* sets of positive integers is denumerable."

· 4 ·

How can the set of all finite sets of positive integers be enumerated? What strategy would you use to get free?

"What about the set of *all* sets of positive integers—all infinite sets as well as all finite ones?" asked Annabelle. "Is that set denumerable or not? Or is the answer unknown?"

"Ah!" said the Sorcerer. "That set is non-denumerable— that is Cantor's great discovery!"

"No one has yet found a way of enumerating that set?" asked Alexander.

"No one has, and no one ever will, because it is logically impossible to enumerate that set."

"How is that known?" asked Annabelle.

"Well, let's first look at it this way: Imagine a book with denumerably many pages—page 1, page 2, page 3, . . . page n, . . . On each page is written down a description of a set of positive integers. You own the book. If *every* set of positive integers is listed in the book, you win a grand prize. But I tell you that you can't win the prize because I can describe a set of positive integers that couldn't possibly be described on any page of the book."

• 5 •

Describe a set of positive integers that is definitely not described on any page of the book.

"And so you see," said the Sorcerer, after explaining the
solution of the last problem, "that the set of all sets of positive
integers is non-denumerable. It is larger than the set of positive
integers."

"You haven't shown that," said Annabelle. "You have shown
that the set of all sets of positive integers—is there a name for
this set?"

"Yes," said the Sorcerer. "For any set A, the set of all subsets
of A is called the *power set of A* and is denoted P(A). We can
use N for the set of positive integers, and so the set of all
subsets of N—which is the set of all sets of positive integers—is
thus the power set of N and is denoted P(N).

"All right," said Annabelle. "Anyway, you have indeed
shown that P(N) is non-denumerable—that P(N) cannot be put
into a 1 to 1 correspondence with N, and of course P(N) is
infinite, but to conclude that P(N) is therefore *larger* than N is
unwarranted, since you haven't shown that N can be put into
a 1 to 1 correspondence with some *subset* of P(N). Don't you
have to do that in order to complete your argument?"

"We have already put N into a 1 to 1 correspondence with
a subset of P(N)," replied the Sorcerer.

• 6 •

When was that done?

After the Sorcerer reminded Annabelle of Problem 4, one of
the two (I forget whether it was Annabelle or Alexander)

raised a question: "We have seen that the set of all *finite* sets of positive integers is denumerable; hence it can be enumerated in some infinite sequence $S_1, S_2, \ldots S_n, \ldots$ Why can't we apply Cantor's argument and get a set S different from all the sets $S_1, S_2, \ldots S_n, \ldots$? Doesn't this create a paradox?"

· 7 ·

Does it create a paradox?

"I have a question," said Alexander after the last matter had been settled. "We know that the set of all *finite* sets of positive integers is denumerable. What about the set of all *infinite* sets of positive integers? Is that set denumerable or not?"

· 8 ·

What is the answer to Alexander's question?

"Another question. We have seen that P(N) is larger than N. Is there a set larger than P(N)?" asked Annabelle.

"Why certainly," replied the Sorcerer. "The fact that P(N) is larger than N is only a special case of Cantor's theorem, which is:

Theorem C (Cantor's theorem). For *any* set A, the set P(A) of all subsets of A is larger than A.

"The proof of Cantor's theorem," said the Sorcerer, "is not substantially different from the proof I have given you for the special case when A is the set N of positive integers. The idea behind the proof has been nicely illustrated by Smullyan in the following problem: In a certain universe, every set of inhabitants forms a club. The registrar of the universe would like to name each club after an inhabitant in such a way that no two clubs are named after the same inhabitant and each inhabitant has a club named after him. It is not necessary that an inhabitant be a member of the club named after him. Now, for a universe with only finitely many inhabitants, this is clearly impossible, for there are then more clubs than inhabitants (if n is the number of inhabitants, then 2^n is the number of clubs). But this particular universe has infinitely many inhabitants, and so the registrar sees no reason why this should not be possible. However, every scheme he has ever tried has failed—there are always clubs left over. Why is the registrar's scheme impossible to carry out?"

· 9 ·

Explain why the registrar's scheme is impossible and how this is related to Cantor's theorem.

"As a consequence of Cantor's theorem," said the Sorcerer after Annabelle and Alexander understood the proof, "there must be an infinite number of different-size infinities, for we can start with N, the set of positive integers, then we have its power set $P(N)$—the set of all subsets of N—which is larger than N, but then again by Cantor's theorem the power set of *this* new set—in other words $P(P(N))$—is larger than $P(N)$, and then the set $P(P(P(N)))$ is larger still, and so forth. Thus, for *any* set, there is a larger set, and so the sizes of sets are endless."

Solutions

1.   For each n, there are only finitely many possibilities for a pair whose highest number is n—there are exactly n such possi-

bilities in fact. Thus there is only one possibility for the pair whose highest number is 1—namely (1, 1). There are two possibilities for the pair whose highest number is 2—namely (1, 2) and (2, 2). Then, there are the three possibilities for the pair whose highest number is 3—(1, 3), (2, 3), and (3, 3), and so forth. And so we enumerate them in the order (1, 1), (1, 2), (2, 2), (1, 3), (2, 3), (3, 3), (1, 4), (2, 4), (3, 4), (4, 4), . . . (1, n), (2, n), . . . (n − 1, n), (n, n), . . .

2.   In this case, it might take us roughly twice as long to get out, but we still get out sooner or later by enumerating the ordered pairs in the order (1, 1), (1, 2), (2, 1), (2, 2), (1, 3), (2, 3), (3, 3), (3, 2), (3, 1), . . . (1, n), (2, n) . . . (n − 1, n), (n, n), (n, n − 1), . . . (n, 2), (n, 1), (1, n + 1), . . .

3.   This is really the same problem as the last, except that we have one integer *above* the other (as the numerator) instead of to the left of the other. Thus, we can enumerate the positive fractions in the order $\frac{1}{1}$, $\frac{1}{2}$, $\frac{2}{2}$, $\frac{1}{3}$, $\frac{2}{3}$, $\frac{3}{3}$, $\frac{3}{2}$, $\frac{3}{1}$, $\frac{1}{4}$, $\frac{2}{4}$, $\frac{3}{4}$, $\frac{4}{4}$, $\frac{4}{3}$, $\frac{4}{2}$, $\frac{4}{1}$, . . . Of course, we could get out a bit sooner by not naming duplications, such as $\frac{2}{2}$ (which is really $\frac{1}{1}$), and $\frac{3}{3}$, and $\frac{2}{4}$ (which duplicates $\frac{1}{2}$), and so forth.

4.   A set whose members are elements a_1, a_2, . . . a_n is written $\{a_1, a_2, . . . a_n\}$. Now, there is only one set whose highest number is 1, namely $\{1\}$. There are two sets whose highest number is 2, namely $\{1, 2\}$ and $\{2\}$. There are four sets whose highest number is 3, namely $\{3\}$, $\{1, 3\}$, $\{2, 3\}$, $\{1, 2, 3\}$. In general, for any positive integer n, there are 2^{n-1} sets whose highest number is n. The reason is this: For any n, there are 2^n subsets of the set of positive integers from 1 to n (and this includes the empty set!). And so any set whose highest number is n consists of n together with some subset of the integers from 1 to n − 1, and there are 2^{n-1} such subsets.

Anyhow, the important thing is that for each n, there are only *finitely* many sets of positive integers whose highest member is n. And so, I first name the empty set. Then, I name the set whose highest number is 1. Then I go through the sets whose highest number is 2 (the order doesn't really matter), then the sets whose highest number is 3, and so forth.

5. Given any number n, either n belongs to the set listed on page n or it doesn't. We let S be the set of all numbers n such that n does *not* belong to the set listed on page n. For each n we let S_n be the set listed on page n. Our definition of S is such that for every n, the set S must be different from S_n, because n belongs to S if and only if n *doesn't* belong to S_n. This means that either n is in S but not in S_n, or n is not in S but in S_n. In either case, S must be different from S_n, because one of those two sets contains n and the other doesn't.

To give a more concrete idea of the construction of the set S, suppose the sets listed on the first ten pages are as follows:

Page 1—Set of all even numbers

Page 2—Set of all (positive whole) numbers

Page 3—The empty set

Page 4—Set of all numbers greater than 100

Page 5—Set of all numbers less than 58

Page 6—Set of all prime numbers

Page 7—Set of all numbers that are not prime

Page 8—Set of all numbers divisible by 4

Page 9—Set of all numbers divisible by 7

Page 10—Set of all numbers divisible by 5

I have just listed the first ten sets at random. Now, what will S look like as far as the first ten numbers are concerned? Well, what about 1; should S contain 1? Is 1 a member of the set listed on page 1; that is, is 1 an even number? No, it is not, so 1 doesn't belong to S_1; so we put 1 in S. What about 2? Well, 2 certainly is in S_2 (2 is a positive whole number), and so we *don't* allow 2 in S. The number 3 is certainly not in S_3 (no number is in the empty set), and so we take 3 as a member of

S. Now, 4 is not in S_4 (4 is not greater than 100), so 4 is in S. We let the reader check the next six cases: 5 is in S, since 5 is not in S_5; 6 is not in S_6 (6 isn't prime), so 6 is in S; 7 is not in S_7, so 7 is in S, 8 is in S_8, so 8 is not in S; 9 isn't in S_9, so 9 gets put in S; 10 is in S_{10} (10 is divisible by 5), so 10 is left out of S. Thus in listing the entries of S, let us put the number n in the nth place if n is in S, and let us put a blank in the nth place to signify that n is definitely left out of S. Then, the first ten places of our list look like this: 1, —, 3, 4, 5, 6, 7, —, 9, — .... We now see that our set S is different from S_1 since it contains 1 and S_1 doesn't. Also, S is different from S_2 because it *doesn't* contain 2 and S_2 does. And so you see, for each n, either S contains n and S_n doesn't, or S doesn't contain n and S_n does, so S cannot equal S_n. Thus, the set S is different from every one of the sets listed in the book.

Of course, we don't really need the book for this argument to hold true. The whole point is that given *any* enumeration S_1, S_2, ... S_n, ... of sets of positive integers, there exists a set S of positive integers that is different from every one of the S_n's—namely, the set of all n such that n doesn't belong to S_n. Thus, the infinite sequence S_1, S_2, ... S_n, ... fails to catch *every* set of positive integers, since the set S has been left out. And so, the set of all sets of positive integers is not denumerable.

6. In Problem 4 we showed that the set of all *finite* subsets of N is denumerable, and thus N can be put into a 1 to 1 correspondence with the set of all finite subsets of N. Obviously, the set F of all finite subsets of N is a subset of the set of *all* subsets of N—thus F is a subset of P(N) and N can be put into a 1 to 1 correspondence with F.

7. Of course not! The set S is indeed different from each of the finite sets S_n, but all that means is that the set S is not finite.

8.   We know that the set of all finite sets of positive integers is denumerable; hence it can be enumerated in some infinite sequence $F_1, F_2, \ldots F_n \ldots$. That is, to each n we can correspond some finite set F_n of positive integers, and the correspondence is such that every finite set of positive integers is some F_n or other. Now, suppose that the set of all *infinite* sets of positive integers were denumerable. Then it could be enumerated in some infinite sequence $I_1, I_2, \ldots I_n, \ldots$ where for each n, I_n is the infinite set corresponding to the integer n. But then, we could enumerate *all* sets of integers—finite and infinite—in the order $F_1, I_1, F_2, I_2, \ldots F_n, I_n, \ldots$ which is contrary to the fact that the set of all sets of positive integers is non-denumerable.

9.   Suppose the registrar's scheme could be carried out. Then we get a contradiction as follows: Call an inhabitant *sociable* if he belongs to the club named after him and *unsociable* if he doesn't. Since in this universe every set of inhabitants constitutes a club, then the set of all unsociable inhabitants constitutes a club. This club is named after someone—say *John*. Thus, every member of John's Club is unsociable and every unsociable inhabitant belongs to John's Club. Is John sociable or not? Either way, we get a contradiction: Suppose John is sociable. This means that John belongs to John's Club, but only unsociable people belong to John's Club, so this is out. On the other hand, suppose that John is unsociable. Since every unsociable inhabitant belongs to John's Club, then John, being unsociable, would have to belong to John's Club, which makes John sociable. And so whether John is sociable or not, we get a contradiction. Thus, the registrar's scheme cannot be carried out.

The relation of this problem to Cantor's theorem should be obvious—it is only a special case of Cantor's theorem. Instead of a universe of people, consider an arbitrary set A. Suppose we have a 1 to 1 correspondence between A and the set P(A) of all subsets of A. We get the following contradiction: Each element x of A corresponds to one and only one subset of A, which we might call x's set. Now, let S be the set of all elements x of A such that x does *not* belong to x's set. (In application to the problem above, S is the set of unsociable inhabitants.) Some element b of A corresponds to this set S, and so b's set is the set of all x's having the property that x doesn't belong to x's set. If b belongs to b's set, then b is one of the elements having the property of not belonging to b's set, which is a contradiction. If b doesn't belong to b's set, then b has the property of not belonging to b's set; but every element having this property belongs to b's set, and so b must then belong to b's set, and we again have a contradiction. This proves that there exists no 1 to 1 correspondence between A and its power set P(A).

Of course, A can be put into a 1 to 1 correspondence with some *subset* of P(A) as follows: For any element x, by {x} is meant the set whose only element is x. (Such a set {x} is called a *unit* set or a *singleton*. It has only one element, regardless of how many elements x itself may have.) Well then, we can let each element x of A correspond to the singleton {x}. This correspondence is obviously 1 to 1, and of course {x} is a subset of A (since every element of {x}, of which there is only one—x itself—is an element of A). And so A is thus in a 1 to 1 correspondence with a set consisting of *some* of the elements of P(A).

Since A can be put into a 1 to 1 correspondence with a subset of P(A) and A cannot be put into a 1 to 1 correspondence with all of P(A) (as we have shown), then by definition, P(A) is larger than A. This proves Cantor's theorem.

20

BUT SOME PARADOXES ARISE!

"SOMETHING has been bothering me," said Alexander, on the next visit. "Cantor has proved that for any set A there is a set larger than A—namely P(A). Isn't that right?"

"Yes," said the Sorcerer.

"Well," said Alexander, "suppose we take for A the set of *all* sets. Then by Cantor's theorem there is a set larger than A, but how can there be a set larger than the set of all sets? The power set of A—the set P(A)—is a subset of A, since A contains *all* sets, so how can a subset of A be larger than A itself? I really don't understand!"

"Ah, you have rediscovered a famous paradox that Cantor himself found in 1897," said the Sorcerer. "Later, Bertrand Russell gave a simplified version of Cantor's paradox known as *Russell's paradox* which goes like this: Given any set x, either x is a member of itself or it isn't. For example, a set of chairs is not itself a chair, so no set of chairs is a member of itself. On the other hand, take something like the set of all things conceivable to the human mind. This set is apparently something conceivable to the human mind, hence is apparently a member of itself. Sets that are not members of themselves are called *ordinary* sets, whereas sets that are members of themselves are called *extraordinary* sets. Whether extraordinary sets really exist is open to question, but ordinary sets certainly do exist. Just about all the sets we encounter are ordinary. Now, let B be the set of all ordinary sets. Thus, every ordinary set is a member of B and every member of B is ordinary—no extraordinary sets are in B. Is B a member of itself or isn't it? Either way, we get a contradiction. Suppose B is a member of itself. Then since only ordinary sets are members of B, then B must be one of the ordinary sets; but on the other hand, since B is a member of itself, B must be extraordinary, which is a contradiction. Thus, it is contradictory to assume that B is extraordinary. Now suppose that B is ordinary. Since all ordinary sets belong to B, then B, being ordinary, must belong to B, thus making B extraordinary (since B then belongs to itself), and so it is also contradictory to assume that B is ordinary. This is Russell's famous paradox. It is a simplification of Cantor's paradox in that the notion of *size* is not involved. I will discuss the possible resolutions of Cantor's and Russell's paradoxes later.

"In 1919, Russell gave a popularization of the paradox in terms of the barber of a certain village who shaves all and only those inhabitants of the village who don't shave themselves. Thus, the barber won't shave any inhabitant who shaves himself, but any inhabitant who does not shave himself is shaved by the barber. Does the barber shave himself or doesn't he? If he does, then he is shaving someone (namely himself) who shaves himself, hence violating the rule that he never shaves anyone who shaves himself. If he doesn't shave himself, then he is one of the inhabitants who doesn't shave himself, but he must shave every such inhabitant, and so he must then shave himself, and we again have a contradiction. So does the barber shave himself or doesn't he? How do you solve this paradox?"

"Perhaps the barber was a woman," suggested Annabelle.

"That won't help," said the Sorcerer. "I didn't say that the barber shaved all the *men* in the village who didn't shave themselves, but that the barber shaved all the *inhabitants* of the village who didn't shave themselves."

"Then what is the solution?" asked Annabelle.

"We'll discuss it later," replied the Sorcerer. "First I'd like to tell you some variants of this paradox.

"There is the paradox of Mannoury, about a certain country in which every municipality must have a mayor and no two municipalities may have the same mayor. A mayor may or may not be a resident of the municipality of which he is mayor. A law is passed setting aside a special municipality called Arcadia that is exclusively for nonresident mayors, and by law every nonresident mayor is compelled to reside there. Arcadia, like every other municipality, must have a mayor. Should the mayor of Arcadia reside in Arcadia or not?

"Then there is the ancient dilemma of the crocodile: A certain crocodile has stolen a child. The crocodile promises the child's father to return the child if and only if the father correctly guesses whether the crocodile will return the child or not. What should the crocodile do if the father guesses that the crocodile will not return the child?

"I am reminded of a paradoxical situation that arose with the logician Smullyan when he was once cleverly outwitted by a student. In his introductory logic classes, Smullyan liked to illustrate an essential idea behind Gödel's proof as follows: He would put a penny and a quarter on the table and say to a student: 'You are to make a statement. If the statement is true, then I promise to give you one of the two coins, not saying which one. But if the statement is false, then you get neither coin.' The problem was to find a statement that would force Smullyan to give the quarter (assuming, of course, that Smullyan kept his word)."

• 1 •

For those who don't know this puzzle, what statement would work? (The solution is given below.)

• 2 •

After Annabelle and Alexander had solved problem 1, the Sorcerer said "Now, a clever student made a statement which made it impossible for Smullyan to keep his word. What statement could it have been?"

Solutions to Problems 1 and 2

1. A statement that works is: "You will not give me the penny." If the statement were false—if it were false that I won't give you the penny—that would mean that I *do* give you the penny, but then I would give you a coin for a false statement, which I said I wouldn't do. Therefore, the statement can't be false; it must be true. Since it is true, that means that I really won't give you the penny; yet I must give you one of the two coins for making a true statement. Hence, I have no choice other than to give you the quarter.

The reader may wonder how this is related to Gödel's theorem. Well, Smullyan thought of the quarter as representing truth and the penny as representing provability. Then the statement "You will not give me the penny," or "I will not be given the penny" corresponds to Gödel's sentence "I am not provable."

2. A statement that makes it impossible for Smullyan to keep his word is: "You will give me neither coin." Whatever I do, I will have to break my promise. (Incidentally, this story is apocryphal; it never really happened to me! I don't know how the

Sorcerer ever got this idea, but I admit that the story is a good one.)

"I know another amusing incident about Smullyan," said the Sorcerer.

"You seem to have an eerie connection with this character Smullyan," said Annabelle. "You have referred to him several times. Do you know him personally?"

"No, the two of us have never actually met. In a sense, we live in different realities. I have reason to believe that he believes that I don't really exist—that I am only a fictitious character. I believe that he believes that you two don't exist either. Really now, how silly can one get!"

"Perhaps it's Smullyan who doesn't really exist," suggested Alexander.

"I have thought of that possibility," said the Sorcerer, "in which case the stories I have heard about him are only legends. Anyhow, whether it is a legend or the truth, as I heard the matter, Smullyan was once giving a graduate course in axiomatic set theory. During the middle of one of his lectures, a girl student of his came in, apologized for being late, and asked Smullyan whether she could have a copy of the notes. Smullyan replied: 'You can have a copy, provided that you are good!' She then asked, 'Just what does it mean to be good?' Smullyan replied, 'It means not knowing what it means to be good!' This got a general laugh."*

"That's very curious," said Annabelle, "because according to Smullyan's definition of 'good,' once a person has heard the definition, that person can never be good, for he or she then knows what 'good' means—it means not knowing what 'good' means."

---

*This story really is true, but how the Sorcerer, who I *insist* is a fictitious character, ever found it out is beyond me! How does a nonexistent person manage to find out things?—R.S.

"I'm not sure about that," said the Sorcerer. "How can one *know* that 'good' means not knowing what 'good' means? That seems contradictory to me.

"Anyway, let's get back to the other paradoxes. I'd like you to try to answer the following questions."

1. Does the barber shave himself or doesn't he?
2. Does the mayor of Arcadia reside in Arcadia or not?
3. What should the crocodile do when told that it will not return the child?
4. How do you get out of Russell's paradox and Cantor's paradox?

· 3 ·

Before reading further, how would you answer those four questions?

The Sorcerer's Explanations.   "I can't even get the first one," said Annabelle. "I can't see how the barber could either shave himself or not shave himself without contradiction, yet he must do one or the other. I don't know what to think! Is there something wrong with logic?"

"Of course not!" laughed the Sorcerer. "The solution to the barber paradox is so obvious that it's amazing that anyone can be fooled by it! And yet some very intelligent people have been taken in. This reveals an interesting psychological characteristic that's unfortunately only too common."

"Don't keep us in suspense," said Alexander. "What *is* the solution to the barber paradox?"

"I'll give you a hint," said the Sorcerer. "Suppose I told you that a certain man is more than 6 feet tall and also less than 6 feet tall. How would you explain that?"

"I would say that it's impossible," replied Alexander.

"Well, doesn't that give you an idea how to solve the barber paradox?"

"Don't tell me," said Annabelle, "that the solution simply is to deny that there is such a barber?"

"Of course!" said the Sorcerer. "What else? Here I have given you contradictory information about a certain barber and asked you to explain the contradiction. The only explanation is that what I have told you is not true!"

"I never thought of that!" said Annabelle.

"Nor I," said Alexander.

"Exactly!" said the Sorcerer. "And that's the unfortunate psychological characteristic to which I referred—the tendency to believe what one is told."

"Are all the other paradoxes solved the same way?" asked Annabelle.

"More or less," replied the Sorcerer. "Let's look at them one by one. As for the mayor of Arcadia, he cannot possibly obey the law because the law is inconsistent. If the mayor decides to reside in Arcadia, he is breaking the law, since only nonresident mayors may live in Arcadia. If he lives outside Arcadia, he is again breaking the law, because his living outside Arcadia makes him a nonresident mayor, and all nonresident mayors are required to live in Arcadia. Thus it is logically impossible for the mayor to obey the law. This doesn't constitute a paradox; it merely means that the law is inconsistent. As to the puzzle of the crocodile, the answer is simply that the creature is unable to do what it said it would do.

"Now, the Russell and Cantor paradoxes are more serious and disturbing because they show that there is something basically wrong with our way of thinking. What I have in mind is this: Doesn't it seem obvious that given any property, there exists the set of all things having that property?"

"It would certainly seem so!" said Annabelle.

"That seems obvious enough," said Alexander.

"That's the whole trouble!" said the Sorcerer. "This principle—called the *unlimited abstraction principle*—the principle that every property determines the set of all things having that property—that principle indeed seems self-evident, yet it leads to a logical contradiction!"

"How so?" asked Annabelle.

"It leads to the Russell paradox as well as the Cantor paradox. Suppose that it were really true that for any property, there exists the set of all things having that property. Well, take the property of being an ordinary set. Then there must exist the set B of all ordinary sets, and we get the Russell paradox: The set B can be neither a member of itself nor not a member of itself without presenting a contradiction. Thus, the unlimited abstraction principle leads to Russell's paradox. It also leads to Cantor's paradox, since we can consider the property of being a set, and so there is then the set of all sets, and this set is on the one hand as large as any set can be, but on the other hand, for any set there is a larger set (by Cantor's theorem), and so there must be a set larger than the set of all sets, which is absurd. Thus, the fallacy of Cantor's paradox is that there is such a thing as the set of all sets, and the fallacy of Russell's paradox is that there is such a thing as the set of all ordinary sets. These two sets simply cannot exist. Yet, the unlimited abstraction principle, which seems so obvious, leads to these paradoxes, and hence cannot be true. The fact that it *seems* true is what I meant when I said that there was something basically wrong with our naïve way of thinking about sets.

"The discovery of these paradoxes was at first very disturbing, since it seemed to threaten that mathematics might be inconsistent with logic. A reconstruction of the foundations of mathematics was necessary, which I will tell you about next time."

21

RESOLUTIONS

"You were going to tell us how the rise of the paradoxes led to the reconstruction of the foundations of mathematics," said Alexander on the next visit.

"Yes," said the Sorcerer. "At the time, the most comprehensive system of mathematical foundations then known was the system of Gottlob Frege. Its purpose was to derive all of mathematics from a few basic principles of logic and sets. In addition to certain axioms of logic, he took just one axiom for sets—the unlimited abstraction principle that every property determines a set—the set of all things having that property. From just this one axiom of set theory, Frege could derive all the sets that mathematics needed. To begin with, we can take some property that doesn't hold for anything, such as the property of something not being equal to itself, and then we have the set of all things having that property, and this is the empty set (since no thing has that property) and is denoted $\emptyset$. Next, given any entities x and y, we can form the set of all things having the property of either being identical to x or identical to y. This set—denoted {x,y}—has x and y as elements and no others. This also holds if x and y happen to be the same entity, in which case the set {x,y} is simply the singleton {x}—the set whose only element is x.

"And so, we now have the empty set $\varnothing$, and, having this, we then have the set $\{\varnothing\}$ whose only member is the empty set $\varnothing$. This is not to be confused with the empty set itself, since the empty set has no members, whereas the set $\{\varnothing\}$ has a member—namely, $\varnothing$. Having the set $\{\varnothing\}$, we can then form the set $\{\{\varnothing\}\}$ whose only member is $\{\varnothing\}$, then the set $\{\{\{\varnothing\}\}\}$, then the set $\{\{\{\{\varnothing\}\}\}\}$, and so forth, thus getting infinitely many sets! These sets can serve as the natural numbers, which was later realized by Zermelo, who took 0 to be the empty set, then 1 to be $\{\varnothing\}$, then 2 to be $\{1\}$ (which is $\{\{\varnothing\}\}$), 3 to be $\{2\}$, and so forth. It is also possible to derive from Frege's unlimited abstraction principle the *set* of all natural numbers. This set is usually denoted by the symbol ω.

"Next, given any set A, we can talk about the property of being a *subset* of A, and so by Frege's principle there then exists the set of all things having that property—in other words, there exists the set of all subsets of A, and this is the *power* set P(A) of A, which plays such a fundamental role in Cantor's work.

"Next, we can talk about the property of being an element of at least one element of A, and so by Frege's principle, there is the set of all elements of all elements of A, and this set is called the *union* of A and denoted UA. (For example, if A is a set of clubs, then UA is the set of all the *people* in the clubs.)

"Yes, from Frege's unlimited abstraction principle, one neatly gets all the sets one needs to do classical mathematics. There is only one trouble with Frege's system; it is inconsistent! From the unlimited abstraction principle one can get the set of all ordinary sets, which gives Russell's paradox, and also the set of all sets, which gives Cantor's paradox. Sadly enough, just as Frege's monumental work was about to be published, Frege got a letter from Russell explaining that his system was inconsistent, and gave the proof, using Russell's paradox. Frege acknowledged the correctness of Russell's proof and was extremely upset, feeling that his life's work had collapsed. His upset was really needless, since the inconsistency of his system could be corrected, and his work contained an enormous number of the basic ideas used by Russell and others later on. Indeed, Russell had the highest respect for Frege. In his book *The Principles of Mathematics*, written in 1902, Russell says the following about Frege."

> The work of Frege, which appears to be far less known than it deserves, contains many of the doctrines set forth in Parts I and II of the present work, and where it differs from the views which I have advocated, the differences demand discussion. Frege's work abounds in subtle distinctions, and avoids all the usual fallacies which beset writers on Logic. . . . In what follows, I shall try briefly to expound Frege's theories on the most important points, and to explain my grounds for differing where I do differ. But the points of disagreement are very few and slight compared to those of agreement.

"You said," said Annabelle, "that the inconsistency of Frege's system could be corrected. How?"

"That's where the work of reconstruction came in," said the Sorcerer, "and this was done along two main lines. The first was done by Whitehead and Russell in their monumental three-volume work *Principia Mathematica*. The second line was adopted by Zermelo in what is known as *Zermelo set theory*, which was later amplified by Fraenkel to produce what is known as *Zermelo-Fraenkel set theory*—abbreviated Z.F.— and is one of the major mathematical systems in use today. The system of Whitehead and Russell, though free from inconsistency with reasonable certainty, was a relatively complicated affair and is not in general use now, so I would rather tell you of the lines followed by Zermelo.

"Zermelo's basic idea was to replace Frege's unlimited abstraction principle, which leads to an inconsistency, by what is called the *limited* abstraction principle, or *separation* principle, which is this: Given any property *and given any set A*, there exists the set of all elements *of the set A* that have the property. Thus, we cannot speak of the set of *all* x's having the property, as Frege did, but we can speak of the set of all x's *in A* that have the property. The reason why this limited abstraction principle of Zermelo is sometimes called the *separation* principle is that given any set A, a property *separates* those elements of A that have the property from those elements of A that do not. Now, this separation principle has never been known to lead to any contradiction and appears unlikely ever to do so. It is indeed a principle in common use by the everyday mathematician who speaks, for example, of the set of all *numbers* having a given property, or if he is doing geometry, he might speak of the set of all *points on a plane* having a given property. He doesn't speak of the set of all *things* having a given property; the 'things' come from some set A whose existence has already been established.

"If we use Zermelo's separation principle instead of Frege's unlimited abstraction principle, Russell's paradox disappears: We can no longer form the set of *all* ordinary sets, but given a set A in advance, we can form the set B of all ordinary elements *of the set A*. (We recall that an ordinary set is a set that is not a member of itself.) This leads to no paradox, but merely to the conclusion that B, though a subset of A, cannot be a member of A."

"What's the difference between being a subset and being a member?" asked Alexander.

"To say that a set X is a *member* of a set Y means that Y is a bunch of things and that X is one of them. To say that X is a *subset* of Y is to say not that X itself is a member of Y, but that all *members* of X are also members of Y. For example, suppose

X is the set of male humans on this planet and Y is the set of all humans on this planet. Surely, the set X of all human males is not a member of Y; it itself is certainly not a human, but all its *members* are members of Y—every human male is also a human. Or again, the set of chairs in this house is a *subset* of the set of pieces of furniture of this house, but it is hardly a *member* of this set—it itself is not a piece of furniture. Or again, the set E of all *even* positive integers is a subset of the set N of all positive integers, but E is certainly not a *member* of N; E itself is not a single positive integer."

"I understand," said Alexander.

"All right then, can you see why the set B of all ordinary members of A, though obviously a subset of A, cannot be a member of A?"

· 1 ·

Why is this so?

"Thus the Russell paradox cannot be reconstructed in Zermelo set theory. Nor can the Cantor paradox, since there is no way of proving in Zermelo set theory that there exists the set of all sets. In fact, it can be proved from the separation principle that there is no such thing as the set of all sets. Do you see how?"

· 2 ·

How can this be proved?

· 3 ·

Here is an analogous problem. Suppose you are told that a certain barber shaves all the inhabitants of the town of Podunk who don't shave themselves and that he never shaves any inhabitant of Podunk who does shave himself. Does that necessarily lead to a contradiction?

"Now then," said the Sorcerer, "as a price for having given up Frege's *unlimited* abstraction principle, Zermelo had to take the existence of the sets $\varnothing$, {x,y}, P(A), UA as separate axioms, and he also had to take as an axiom the existence of the set of all natural numbers—this is the so-called *axiom of infinity*. And so the axioms of Zermelo's system are: (1) the separation principle; (2) existence of the empty set $\varnothing$; (3) for any sets x and y, there exists the set {x,y} whose members are just x and y; (4) for any set A, there exists its power set P(A); (5) for any set A, there exists its union UA; (6) the axiom of infinity.

"That is Zermelo's entire system. Much later (in the twenties) Abraham Fraenkel added what turned out to be an extremely powerful axiom called the *axiom of replacement*, which is roughly that, given any set A, one can form a new set by replacing each element of A by any element whatsoever, with the understanding that two or more elements can be replaced by the same element, but no element may be replaced by more than one element (and thus the size of the set A wouldn't be increased).

"That is the famous system Z.F.—Zermelo-Fraenkel set theory. It is in widespread use today, and it is amazing that the whole of classical mathematics—number theory, algebra, calculus, topology, etc.—can be derived from just these few axioms of set theory! I hope one day to give you an idea of how can be done."

Solutions

1. Let B be the set of all ordinary elements of A. Thus B consists of all and only those x's in A such that x is ordinary. Thus, for every x *which happens to be in A*, x is in B if and only if x is ordinary. (This is very different from saying that for any x *whatever*, x is in B if and only if x is ordinary. This latter would say that B is the set of all x's whatever such that x is ordinary, whereas B is only the set of all x's *in A* that are ordinary.) Now, if B were an element of A, then B would be one of the x's such that x is in B if and only if x is ordinary—in other words, B would be in B if and only if B were ordinary; but this is absurd, since to say that B is ordinary is to say that B is *not* in B! Thus, the assumption that B is an element of A leads to a contradiction (similar to that of the Russell paradox), but the contradiction is avoided by B being outside (not a member of) A.

2. This follows from Problem 1: Given any set A, it fails to contain the set B of all ordinary elements of A; thus some set is outside A. Hence not all sets are members of A, and so no set A is the set of all sets.

3. No, this leads to no contradiction, since you were not told that the barber himself was an inhabitant of Podunk! The conclusion is simply that the barber is not a resident of Podunk, for if he were, he could neither shave himself nor not shave himself without contradicting the given conditions. Living outside of Podunk, however, he can either shave himself or not shave himself without contradiction, since nothing was told you about the people *outside* of Podunk whom he shaved or didn't shave! (I hope you see the analogy to Problem 1!)

22

THE CONTINUUM PROBLEM

"One thing I would like to know," said Annabelle. "We know that the set P(N) of all sets of positive integers is larger than the set N of positive integers. Is there a set A that is larger than N but smaller than P(N)? In other words, is there some set of size intermediate between the sizes of N and P(N) or is P(N) the set of next larger size than N?"

"Ah!" said the Sorcerer, "wouldn't you like to know indeed? And wouldn't I like to know, and wouldn't the whole mathematical world like to know! This was the basic question raised by Cantor and the answer is not known to this day! Cantor *conjectured* that there is no set whose size is intermediate between that of N and that of P(N), and this conjecture is known as the *continuum hypothesis*. But it is only a hypothesis or conjecture; it has not been either proved or disproved to this day. Cantor conjectured the more general hypothesis that for *no* infinite set A is there a set of size intermediate between that of A and that of P(A). This conjecture is known as the *generalized continuum hypothesis*. But again, it is only a conjecture; no one has yet proved or disproved it. I personally regard this unsolved problem as *the* grand unsolved problem—the most interesting unsolved problem in all of mathematics. Many mathematicians and logicians feel as I do about this."

"Why the word 'continuum'?" asked Alexander.

"It so happens that the set P(N) can be put into a 1 to 1 correspondence with the set of points on an infinite straight line, and a straight line is sometimes called a continuum. Thus, P(N) is said to be of the same size as the continuum. And so, the question is whether there is a set larger than N but smaller than the continuum."

"What are the prospects of solving the continuum problem?" asked Annabelle.

"That's very difficult to say," replied the Sorcerer. "In 1939, Gödel proved that if we take the system of Zermelo-Fraenkel—which is one of the most powerful systems of mathematics yet known—the continuum hypothesis cannot be disproved in it. And in 1963, Paul Cohen proved that the continuum hypothesis can never be *proved* from these axioms. And so the continuum hypothesis is *independent* of the present-day axioms of set theory."

"Does this mean that the continuum hypothesis is neither true nor false, but depends merely on what axiom system you take?" asked Annabelle.

"That's a highly controversial question," replied the Sorcerer. "There are those called *formalists* who regard the continuum hypothesis as neither true nor false, but entirely dependent on what axiom system you take, since we can add either the continuum hypothesis or its negation to the axioms of set theory and have a consistent system in either case—assuming, of course, that the axioms of set theory are themselves consistent, of which there is little doubt. And so, the formalist doesn't believe that the continuum hypothesis is in itself either true or false, but merely dependent on what axiom system one takes. At the other extreme are the so-called mathematical *realists*, or *Platonists*—of which I am definitely one—who believe that of course the continuum hypothesis is either true or false, but we don't know which. We believe that we don't yet know enough about sets to answer the question, but this doesn't mean that the question has no answer!

"The formalist position really strikes me as most strange! Physicists and engineers certainly don't think like that. Suppose a corps of engineers has a bridge built and the next day the army is going to march over it. The engineers want to know whether the bridge will support the weight or whether it will collapse. It certainly won't do them one bit of good to be told: 'Well, in some axiom systems it can be proved that the bridge will hold, and in other axiom systems it can be proved that the bridge will fall.' The engineers want to know whether the bridge will *really* hold or not! And to me (and other Platonists) the situation is the same with the continuum hypothesis: is there a set intermediate in size between N and P(N) or isn't there? If the formalist is going to tell me that in one axiom system there is and in another there isn't, I would reply that that does me no good unless I know which of those two axiom systems is the *correct* one! But to the formalist, the very notion of *correctness*, other than mere consistency, is either meaningless or is itself dependent on which axiom system is taken. Thus the deadlock between the formalist and the Platonist is pretty hopeless! I don't think that either side can budge the other one iota!"

"I didn't realize," said Annabelle, "that there was that amount of controversy in a field like mathematics! I thought that field was cut and dried and there was no room for differences of opinion."

"The differences of opinion are not so much in mathematics as in the *foundations* of mathematics. And the subject of mathematical foundations comes quite close to philosophy, in which, of course, there is great difference of opinion," replied the Sorcerer.

"What about Gödel and Paul Cohen themselves?" asked Alexander. "Are they formalists or Platonists?"

"I'm not sure about Cohen," replied the Sorcerer. "In fact, I'm not sure that Cohen has made up his mind about the matter, though I suspect he is somewhat close to formalism— but please don't quote me on this, because I really don't know. Now, with Gödel, he was emphatically a Platonist! He explicitly stated that what was needed was to find new axioms of set theory which are as self-evidently true as the present ones and which would be strong enough to settle the continuum hypothesis one way or the other. And he predicted that one day such axioms would be found and when they were, Cantor's continuum hypothesis would be seen to be *false!* Yes, Gödel proved that although the continuum hypothesis—even the generalized continuum hypothesis—could never be disproved from the present-day axioms of set theory, it is nevertheless false.

"Well," concluded the Sorcerer, "so far Gödel's hope has not been realized. No new self-evidently true axioms have yet been discovered that can settle the question. Will they ever be discovered? Who knows? If and when they are, it will certainly be a glorious day!"

PART VII

HYPERGAME, PARADOXES, AND A STORY

23
HYPERGAME

"Do you know the paradox *hypergame?*" asked the Sorcerer one day.

Neither Annabelle nor Alexander had heard of it.

"It is a lovely paradox created in the eighties by the mathematician William Zwicker. Besides being a delightful paradox in its own right, it leads to a totally new proof of Cantor's theorem."

"That sounds interesting!" said Annabelle.

"Well, first for the paradox," said the Sorcerer. "We will be discussing games played by just two people. Call a game *normal* if it has to terminate in a finite number of moves. An obvious example is tic-tac-toe; it must end in at most nine moves. Chess is also a normal game; the fifty-move rule ensures that the game cannot go on forever. Checkers is also a normal game. Every card game that I know is normal. Possibly chess, if played on an infinite board, might be nonnormal.

"Now, here is *hypergame:* The first move in hypergame is to declare what *normal* game should be played. Suppose, for example, that one of you is playing against me and I have the first move. Then I must declare what normal game should be played. I might say: 'Let's play chess,' in which case you make the first move in chess and we keep playing until the chess game is terminated. Or, instead, I might say: 'Let's play checkers.' Then you make the first move in checkers, and we continue playing until the checker game is terminated. Or I might say: 'Let's play tic-tac-toe'—I can choose *any* normal game I like. But I am not allowed to choose a game that is not normal; I must choose a *normal* game.

"Now the problem is this: Is hypergame normal or not?"

The two thought about this for a while and came to the conclusion that the game must be normal.

"Why?" asked the Sorcerer.

"Because," they explained, "whatever normal game is chosen, that game must eventually terminate, since it is normal. This terminates the hypergame that is played. And so regardless of what normal game is chosen, the process has got to terminate. Thus, hypergame must be normal."

"So far, so good," said the Sorcerer, "but then a problem arises. Now that it is established that hypergame is normal, and since I may select *any* normal game on my first move, then I can say: 'Let's play hypergame.' Then you can say: 'OK, let's play hypergame.' And then I can say: 'OK, let's play hypergame,' and this process can go on indefinitely. Thus, hypergame does *not* have to terminate, which means that hypergame is not normal after all! And yet, you have proved that it is normal! This is a paradox."

Neither Annabelle nor Alexander could solve it.

"The whole point," said the Sorcerer, "is that the general notion of a game is not well defined. Given a set S of well-defined games, one can indeed define a hypergame *of that set* S, but this hypergame cannot itself be one of the games of S.

"Now, someone—I think Hegel—once defined a paradox as a truth standing on its head. Very often what first comes out as a paradox gets modified and leads to an important truth. And so it is with Zwicker's paradox of hypergame. A modification of the argument establishes an interesting theorem that yields a completely new proof of Cantor's theorem.

"Recall briefly Cantor's proof. We are given a set A and to each element x of A is associated a subset of A denoted S_x. The idea is to construct a subset C of A that is different from S_x for every x. Cantor took C to be the set of all elements x of A such that x doesn't belong to S_x. Now, what Zwicker did was to find an entirely different set Z that is distinct from every one of the sets S_x. It, like Cantor's argument, shows that it is impossible to put A into a 1 to 1 correspondence with the set of all subsets of A, but the new set Z got by Zwicker is entirely different from the set C got by Cantor. Here is what Zwicker did.

"Once the correspondence (that assigns to each x in A the subset S_x) is given, we define a *path* to be any finite or infinite sequence x,y,z, . . . of elements of A such that for each term w of the sequence, either w is the last term, or the next term is an element of S_w. Thus, a *path* is generated as follows: Start with an arbitrary element x of A. If S_x is empty, that's the end of the path. If not, pick some element y of S_x. We then have the two-term sequence (x,y). If S_y is empty, that's the end of the path. If not, pick some element z of S_y, and we then have the three-term sequence (x,y,z). If S_z is empty, that's the end of the path, but if S_z is nonempty, pick some element w and make it the fourth term of the path, and keep on going in this manner until you either come to some S_v that is empty, in which case the path ends, or else keep going without stop, thus generating an infinite path. [For example, if y is an element of S_x and x is an element of S_y, then (x,y,x,y,x,y, . . .) would be an infinite path. Or if x happens to be in S_x, then (x,x,x,x, . . .) would be an infinite path.] Now, given any x, either there does or there doesn't exist an infinite path that starts with x. Now define x to be *normal* if there exists no infinite path starting with x. Thus, if x is normal, then *every* possible path starting with x must terminate. Now let Z be the set of all the normal elements. We then have

Theorem Z—(Zwicker's theorem).   The set Z is different from S_x for every x.

"The proof," said the Sorcerer "is an obvious modification of the argument establishing the paradox in hypergame."

• 1 •

Give the proof of Zwicker's theorem.

"Notice," said the Sorcerer, "that Zwicker's set Z bears no relation to Cantor's set C. The set of *normal* elements bears no significant relation to the set of x's that don't belong to S_x.

"Cantor's proof relies essentially on the notion of negation. C is the set of all x's such that x does *not* belong to S_x. Zwicker's proof is not based on negation; it is based on the notion of *finiteness* instead."

"It seems to me," said Annabelle, "that the notion of negation is secretly hidden in Zwicker's proof. He defines x to be normal if there exists *no* infinite path starting with x. Isn't that an implicit use of negation?"

"That's a clever observation," said the Sorcerer, "but that use of negation is not really essential. We could have simply defined x to be normal if *all* paths starting with x are finite."

Solution

1.   We are to show that the set Z of normal elements cannot be S_x for any x. Equivalently, we are to show that for no x is it the case that S_x is the set of all normal elements. So suppose

that x is such that S_x is the set of all normal elements. We then get a contradiction as follows:

We will first show that x must be normal. Well, consider any path starting with x. If S_x happens to be empty, the path stops right there with x (since any second term y must be a member of S_x). So we suppose S_x to be nonempty. Then the second term y of the path must be chosen from S_x, hence must be normal (since only normal elements are in S_x). Since y is normal, then every path starting with y must terminate; hence every path starting with (x,y. . .) must terminate, and so x must be normal.

Since x is normal and S_x is the set of *all* normal elements, then x must be a member of S_x. Hence, there is the infinite path (x,x,x, . . .) just like the infinite game ("Let's play hypergame," "Let's play hypergame," "Let's play hypergame," . . .), and we thus get a contradiction.

Thus the set Z of all normal elements must be different from every S_x.

24

PARADOXICAL?

"I RECENTLY THOUGHT of a paradox," said Annabelle. "It concerns your explanation of Cantor's proof in terms of the book of listed sets. You remember? You described a book with denumerably many pages numbered page 1, page 2, . . . and so on, and on each page was written the description of a set of positive integers. The problem was to describe a set that wasn't described on any page of the book. Do you remember?"

"Yes, of course I remember," replied the Sorcerer.

"Very well, and if you remember, the solution you gave was the description *'The set of all n such that n is not a member of the set described on page n.'* "

"That's right," said the Sorcerer.

"Now, my paradox is this: Suppose that that very description occurs on some page of the book—say page 13. Is 13 then a member of that set or not? Just think, we are considering the set S of all numbers n that don't belong to the set described on page n. Thus, for any number n, n belongs to S if and only if n doesn't belong to the set described on page n. In particular, 13 belongs to S if and only if 13 doesn't belong to the set described on page 13, but S *is* the set described on page 13, so we have the absurdity that 13 belongs to S just in case it doesn't belong to S! How can that be?"

258

"That's neat!" said the Sorcerer, with a broad smile. "I like that idea very much!"

"But what is the solution?" pleaded Annabelle.

• 1 •

Before reading further, does the reader have any idea of how to get out of this one?

Solution to Problem 1. "The explanation is this," said the Sorcerer. "Consider the following expression:

(1) The set of all n such that n doesn't belong to the set described on page n.

"Now, if (1) appears on one of the pages—say page 13—then it is not a genuine description of any set; it is what is called a *pseudodescription*."

"Why?" asked Annabelle.

"Because if it were, then it would lead to a contradiction— the very contradiction you so aptly described."

"I'm not sure that explanation is quite satisfactory," said Annabelle.

"Look at it this way," said the Sorcerer. "For a description of a set of numbers to be genuine, it must provide a definite rule—a definite criterion for which numbers are in the set and which numbers are not. If the expression (1) appears on page 13 of the book, then for every n *other than 13*, it tells you whether n is in the set or not. But it does *not* tell you whether 13 is or is not in the set. Now, the following is a genuine description, even if it does appear on page 13.

(2) The set of all n other than 13 such that n doesn't belong to the set described on page n.

"According to this, the number 13 is *not* a member of the set described on page 13. Also the following, if appearing on page 13, is a genuine description.

(3) The set of all n other than 13 such that n doesn't belong to the set described on page n, together with 13.

"This is genuine, because it tells you what to do with 13—13 is to be included in the set. And so (2) and (3) are genuine descriptions, even if one of them appears on page 13, but (1), if it appears on page 13, is not a genuine description. The curious thing is that (1) is genuine *providing it doesn't appear on any page of the book!* If written outside the book, it is genuine—providing, of course, that all the descriptions in the book are genuine."

At this point, the Sorcerer stared out into space for a couple of minutes, lost in thought. When he came to, he said: "You know, you have just given me an idea for a much more baffling paradox! All right, we now consider another book with denumerably many pages and on each page is written either a genuine description or a pseudodescription of a set of positive integers. Unlike the last book, we now allow pseudodescriptions to appear on some of the pages. Now, the following description is surely genuine:

> The set of all n such that the description on page
> n is genuine and n doesn't belong to the set
> described on page n.

"This description must be genuine because it provides a definite rule for which numbers belong to the set S so described and which numbers do not. Consider any number n. The description on page n is either genuine or it isn't. If it isn't, then n is automatically excluded from S. If it is, then the description on page n really describes a set; hence it is really determined whether n is in that set and it is accordingly determined whether n is in S or not. Therefore, the above description is indeed genuine.

"Now, what happens if *that* genuine description is written on page 13? Is 13 then in the set so described or isn't it? Either way, you get a contradiction, as Annabelle has shown us, and we can't get out of it this time by saying that the description is not genuine, for I have just proved to you that it is! Well?"

"You are certainly making life difficult!" said Alexander.

"I realize that," said the Sorcerer, with a mischievous smile.

· 2 ·

Well, how does one get oneself out of *this* one? (Solution is given at the end of the chapter.)

Relation to the Berry Paradox. "Actually," said the Sorcerer, after he gave the solution, "my paradox is closely related to the Berry paradox—it is a sort of Cantorian version of it."

"What is the Berry paradox?" asked Annabelle.

"It is this: as numbers get larger and larger, it takes more and more words to describe them."

"That seems reasonable," said Alexander.

"In fact for any positive integer n, there must be numbers that cannot be described in less than n words."

"I believe that," said Alexander.

"Then for any n, there must be the *smallest* number not describable in less than n words. Right?"

"Of course," said Annabelle.

"All right, now look at the following description.

THE SMALLEST NUMBER NOT DESCRIBABLE
IN LESS THAN ELEVEN WORDS.

"That describes a definite number, doesn't it?" asked the Sorcerer.

The couple agreed.

"Will you please count the number of words in that description," asked the Sorcerer.

They counted the number and realized to their horror that it was 10.

"And so, the smallest number not describable in less than eleven words is describable in ten words. Now, will you *please* explain to me how that can be?" said the Sorcerer.

· 3 ·

Oh, no! How *could* that be?

"All these paradoxes," said the Sorcerer, "remind me of a very delightful story of how the Devil was once outwitted by a clever student of Georg Cantor. Do you know the story of Satan, Cantor, and Infinity?"

Neither Annabelle nor Alexander knew the story.

"Then I will tell it to you on your next visit."

Solutions

1.   This solution is given immediately after the problem.

2.   The explanation is that the very notion of a genuine description is not well defined. One can define a genuine description only within a precisely formulated language, and English is not such a language. The situation is similar to that of *truth*.

As has been shown by the logician Alfred Tarski, for languages of sufficient strength, truth for sentences of the language is not definable within the language. For example, *truth* for English sentences is not definable in English, for if it were, you would get the paradoxical sentence "This sentence is not true."

3.   The solution is really the same as that of the last problem. The notion of a description is not well defined.

25

SATAN, CANTOR, AND INFINITY

Here is the story as told by the Sorcerer.

"I have been having a lot of fun with some of our victims," said Satan to Beelzebub, as he rubbed his hands in glee. "In each case I tell the victim that I am thinking of one and only one object out of an infinite set of objects. Each day the victim is allowed one and only one guess as to what the object could be. If and when he guesses it, he goes free. This is the general format of these tests. In some cases the victim has been clever enough to devise a strategy to win his freedom, in other cases not. Well, tomorrow I am expecting a new victim and I will arrange matters so he can *never* go free!"

"What will you do?" asked Beelzebub.

"I have written down the name of a *set* of positive whole numbers. Each day the victim will be allowed to name one and only one set, and if he ever names my set, he can go free. But he will *never* go free!" said Satan, shrieking with delight.

"Why?" asked Beelzebub.

"Well, just look at what I have written!"

The set of all numbers n such that n does not belong to the set named on the nth day.

"I don't understand!" said Beelzebub.

"I thought you wouldn't, blockhead!" said Satan. "He can't possibly name *my* set on any day, because for each positive integer n, the set he names on the nth day is different from my set, since one of these two sets contains the number n and the other doesn't! It's as simple as that!"

"That sounds like fun!" said Beelzebub.

Well, it so happened that the next victim was a prize student of Georg Cantor! He not only knew his mathematics of infinity perfectly but was also an expert in semantics and law. In fact, he had planned to go into law before he fell under Cantor's magnetic sway and decided instead to go into logic and mathematics.

"Before I sign any contract," said the student to Satan, "I want to be sure that I'm absolutely clear as to the terms."

"I've already told you," said Satan, "that I have written down the name of some set of positive integers and it is right here in this envelope with my royal seal. Each day you are allowed to name one and only one set. If and when you name the set written here, you go free. It's as simple as that!"

"I already understood that," replied the student, "but there are several questions that need to be answered. First, suppose that on a given day I name the same set that you have written, but my *description* of the set is different from yours. Any set can be described in many different ways. For example, suppose you have written, 'The set whose only member is the number 2,' but on some day I say, 'The set of all even prime numbers.' Now, the two sets are really the same, since 2 is the *only* even prime number; yet the descriptions are different. What happens then?"

"Oh, in that case you win," replied Satan. "I do not demand that our *descriptions* be the same but only that they describe the same set."

"But that raises a serious problem!" said the student. "It is not always a simple matter to determine whether two descriptions name the same set. Suppose on a certain day I name a set and you reply, 'No, that's not the set I have in mind,' but I have reason to believe that it really *is* the same set and that you have only described it differently. What happens then?"

"In that case," said Satan, "you are allowed to challenge me. Now, a challenge is a very serious matter and you should think very carefully before you make one. It might win you instant freedom, or it might doom you here forever!"

"Just what do you mean by a 'challenge'?" asked the student.

"You challenge me to open the envelope and show you what I have written. If you can prove that the two descriptions—yours and mine—are really of the same set, you win the challenge and go free. But if I can prove that the descriptions name different sets, then you have lost the challenge and your right to name any more sets in the future is canceled. There is then no way you can ever escape. Remember well that after a challenge, you are not allowed to name any more sets."

"That's clear enough," said the student, "but now comes a second point. How do I know that you have really written the name of a set in this envelope?"

"You doubt my word?" asked Satan.

"Oh, not at all; I don't doubt for a moment that you have written something in this envelope which you *believe* to be a genuine description of a set, but it has happened in the history of mathematics that what at first sight appeared to be a genuine description has turned out not to describe any set at all—in other words, that there really is no set answering such a description. Such 'descriptions' are what logicians call *pseudodescriptions*. They appear to describe a set but really don't. Now suppose that at a certain stage, I have reason to suspect that what you have written in the envelope is not a genuine description but only a pseudodescription. What happens then?"

"If on any day you suspect that I have written only a pseudodescription," replied Satan, "then again you may challenge me. I will open the envelope and show you what I have written. If you can prove that it is only a pseudodescription, you win the challenge and go free. But if I can prove that it really is a genuine description, then you lose the challenge and again your right to name any more sets in the future is canceled. I must earnestly remind you that *after a challenge, you may not name any more sets.*"

"That point is now clear," said the student. "One last thing: Are you willing to have it written in the contract that if at any time you violate any of the conditions, then I go free?"

"Yes," replied Satan, "provided that you are willing to have it written that if at any time *you* violate any of the conditions, then you stay here forever."

"Agreed!" said the student.

The contract was then drawn up by Beelzebub and duly executed by both parties.

"Good!" said Satan. "When would you like to begin?"

"Today's as good a day as any," said the student. "Let this be the first day of the test."

"Very well, then, name a set of positive integers!"

"The set of all n such that n does not belong to the set I name on the nth day," said the student. "And now I challenge you to open the envelope."

"Good grief!" cried Satan. "I never thought of that!"

"Open the envelope!" demanded the student.

Satan had to open the envelope, and of course he had written the same thing.

"So I go free!" said the student.

"Not so fast, young man!" said Satan. "You have not really named a set; you have done just what you accused me of possibly doing; you have given only a *pseudodescription*, not a genuine description!"

"Why?" asked the student.

"Because the assumption that you *have* named a set leads to a logical contradiction: Suppose you have named a set. Then this set is the set you have named on the first day. Now, the number 1 belongs to this set if and only if it doesn't belong to the set named on the first day, but since this set *is* the set named on the first day, then 1 belongs to this set if and only if it doesn't. This is a clear contradiction, and the only way out of the contradiction is that you have not really named a set."

"I'm glad you realize this," said the student, "because by the same token, *you* have failed to name a set."

"Now, just a minute!" said Satan. "The genuineness of *my* description is predicated on the assumption that you name one and only one set each day, as you're supposed to. So far, you have not yet named a set today, so I now command you to name a set."

"Oh, I have no intention of naming any sets today."

"What?" cried Satan, unable to believe his ears.

"It doesn't say anywhere in the contract that I *must* name a set on each day; it says that on each day I am *allowed* to name a set. Well, it so happens that today I don't choose to name any set."

"Oh, really!" shrieked Satan. "You refuse to name a set today, eh? Well I'll *force* you to name a set today, and tomorrow I'll again force you to name a set, and the day after and the day after, and so on throughout all eternity. You have no idea how terrible my methods are!"

"Oh, you can't do that," said the student. "I've already challenged you, and it says quite explicitly in the contract that after a challenge, I'm not allowed to name any more sets."

Epilogue.   Of course, Satan had to set the student free. The student immediately ascended to paradise and embraced his beloved master, Georg Cantor. The two had a delighted chuckle over the entire affair.

"You realize," said Cantor, "that you didn't have to be that elaborate; you didn't have to start the procedure by giving a pseudodescription. You could have started by simply saying: "I challenge you!" After the challenge, you wouldn't have been allowed to name any sets, which would automatically have made Satan's 'description' a mere pseudodescription."

"Oh, I realized that," said the student. "I just thought I'd have a little fun with him."

"You realize," said the Sorcerer to Annabelle and Alexander when he had ended his story, "that Satan used Cantor's famous diagonal device to prove that the power set of a set n has higher cardinality than n. The student rightfully guessed that Satan would try to pull such a Cantorian trick. Several people have asked me whether the expression 'The set of all n such that n does not belong to the set named on day n' is a genuine description or not. The answer is that it is a genuine description if and only if on each day there is one and only one set named on that day. If the student failed to name a set on so much as one day, that would be enough to nullify the meaningfulness of Satan's description. Or if the student named more than one set on a given day, that would also invalidate Satan's description. But if the student named one and only one set on each day, then Satan's description would be perfectly well defined. A curious thing, though, about this description is that after no finite number of days can it be *known* that Satan wrote a genuine description, unless it could somehow be known that the student *would* name one and only one set each day.

"Satan really made a very poor contract! If he had *required* instead of just *allowed* the student to name one and only one set each day and if he had just deleted this business about the student not being allowed to name any more sets after a challenge, he would obviously have won. Had he done that, then it would indeed have been logically impossible for the student ever to name Satan's set. But as the contract stands, a mere challenge on the part of the student disallows him from naming any more sets, which in turn nullifies the genuineness of Satan's 'description.'

"The moral of the story," said the Sorcerer, "is that even fallen angels might benefit from a good course in mathematical logic."

This book was set in Caledonia, a type face designed by W(illiam) A(ddison) Dwiggins (1880–1956) for the Mergenthaler Linotype Company in 1939. Dwiggins chose to call his new type face Caledonia, the Roman name for Scotland, because it was inspired by the Scottish types cast about 1833 by Alexander Wilson & Son, Glasgow type founders. However, there is a calligraphic quality about Caledonia that is totally lacking in the Wilson types.

Dwiggins referred to an even earlier type face for this "liveliness of action"—one cut around 1790 by William Martin for the printer William Bulmer. Caledonia has more weight than the Martin letters, and the bottom finishing strokes (serifs) of the letters are cut straight across, without brackets, to make sharp angles with the upright stems, thus giving a modern-face appearance.

W. A. Dwiggins began an association with the Merganthaler Linotype Company in 1929 and over the next twenty-seven years designed a number of book types, the most interesting of which are Metro, Electra, Caledonia, Eldorado, and Falcon.

Composed, printed, and bound by The Haddon Craftsmen, Scranton, Pennsylvania

Designed by Brooke Zimmer